HEATS OF HYDROGENATION

Experimental and
Computational Hydrogen
Thermochemistry of
Organic Compounds

HEATS OF HYDROGENATION

Experimental and Computational Hydrogen Thermochemistry of Organic Compounds

Donald W. Rogers
Long Island University, USA

 World Scientific

NEW JERSEY • LONDON • SINGAPORE • BEIJING • SHANGHAI • HONG KONG • TAIPEI • CHENNAI

Published by

World Scientific Publishing Co. Pte. Ltd.
5 Toh Tuck Link, Singapore 596224
USA office: 27 Warren Street, Suite 401-402, Hackensack, NJ 07601
UK office: 57 Shelton Street, Covent Garden, London WC2H 9HE

British Library Cataloguing-in-Publication Data
A catalogue record for this book is available from the British Library.

HEATS OF HYDROGENATION
Experimental and Computational Hydrogen Thermochemistry of Organic Compounds

Copyright © 2006 by World Scientific Publishing Co. Pte. Ltd.

All rights reserved. This book, or parts thereof, may not be reproduced in any form or by any means, electronic or mechanical, including photocopying, recording or any information storage and retrieval system now known or to be invented, without written permission from the Publisher.

For photocopying of material in this volume, please pay a copying fee through the Copyright Clearance Center, Inc., 222 Rosewood Drive, Danvers, MA 01923, USA. In this case permission to photocopy is not required from the publisher.

ISBN-13 978-981-256-954-7
ISBN-10 981-256-954-5

Printed in Singapore

Dedicated to the memory of Frank McLafferty
Scholar, Teacher, Friend

He maketh me to lie down in green pastures…

Psalms 23

Preface

Hydrogen thermochemistry consists of obtaining useful thermodynamic information by measuring the heat output upon adding hydrogen across the double or triple bond or bonds of an unsaturated organic molecule. Though far less general in its application than combustion thermochemistry, it is simpler and it produces results of comparable accuracy. Heats of hydrogenation constitute a body of thermochemical information that has had a historical significance far out of proportion to the small number of research groups that have engaged in the work. Early studies (1935–1939) by Kistiakowski and coworkers at Harvard produced some of the most widely quoted results in all of thermochemistry, among them the conjugation energy of 1,3-butadiene and the resonance energy of benzene. After an initial, highly productive period, however, hydrogen thermochemistry was cut short by World War II, and it languished until the late 1950s, largely because chemists favored the vastly more general competing method of oxygen bomb calorimetry.

Recent highly accurate quantum mechanical calculations requiring reference standards of comparable accuracy and the economics of fuels and petrochemicals have brought hydrogen thermochemistry back into contemporary focus. There is at present no critical survey of the scattered literature on experimental methods and results of hydrogen thermochemistry, or the burgeoning field of computational hydrogen thermochemistry. This short monograph is intended to fill that gap. It is divided into three chapters, each covering a distinct part of hydrogen thermochemistry:

1. Practical and historical aspects of experimental determination of the enthalpies of hydrogenation and formation $\Delta_{hyd}H_{298}$ and $\Delta_f H_{298}$ of organic compounds, primarily hydrocarbons, are given in Chapter 1.
2. Chapter 2 consists of a large table containing more than 500 $\Delta_{hyd}H_{298}$ values measured over 70 years of research, now scattered throughout the literature in English and in German. One of the unusual aspects of this literature is that contemporary experimental advances, though they have brought about an increase in sensitivity measured in orders of magnitude, they have not brought about a significant change in accuracy. Thus the older literature contains results that are as valuable today as they were when they were first published. Like all literatures, however, the hydrogen thermochemical literature varies in accuracy and reliability, both of which are commented upon within this table.
3. Contemporary advances in computer hardware and software have brought about a immense increase in our ability to calculate (as distinct from measuring) $\Delta_{hyd}H_{298}$. Modern computational thermochemistry includes various empirical, semiempirical, and quantum mechanical methods for calculating $\Delta_{hyd}H_{298}$ of linear and cyclic alkenes, polyalkenes, alkynes and polyalkynes. There is at present no critical survey of computational methods for determining $\Delta_{hyd}H_{298}$ which Chapter 3 supplies.

Whatever technological problems and successes the future may bring, we can be assured that they will be within the framework of classical thermodynamics.

The author wishes to acknowledge the help and encouragement of colleagues and friends Joel Liebman, Nikita Matsunaga, and Andreas Zavitsas. As always, I express my thanks to Tony Zeru Li for extracting me from the computational problems I have created for myself in writing this book.

Greenwich Village NY   Donald W. Rogers

Contents

Preface vii

1. Hydrogen Thermochemistry
   1.1 Definitions 1
   1.2 History 6
   1.3 Theory and Methodology 8
       1.3.1 Correction to the gaseous state 12
   1.4 Accuracy 15
   1.5 Applications 15
   1.6 Details of Calorimeter Construction 17
   1.7 Design Modifications 21

2. Experimental Results
   2.1 Enthalpies of Hydrogenation 23

3. Computational Thermochemistry
   3.1 Introduction to Computational Thermochemistry 153
   3.2 Molecular Modeling 155
   3.3 Additivity Methods 159
       3.3.1 Bond additivity 159
       3.3.2 Group additivity 163
       3.3.3 The thermochemical database 166
   3.4 Molecular Mechanics 168
   3.5 Molecular Orbital Calculations 172
   3.6 Semiempirical Methods 176
       3.6.1 The Huckel method 176
       3.6.2 Higher semiempirical methods 181

3.7 *Ab initio* Methods 182
    3.7.1 The Gaussian basis set 183
    3.7.2 Post Hartree–Fock methods 185
    3.7.3 Combined or scripted methods 186
    3.7.4 Finding $\Delta_f H_{298}$ and $\Delta_{hyd} H_{298}$ from G3(MP2) results 189
    3.7.5 Variation of $\Delta_{hyd} H_{298}$ with T 191
    3.7.6 Isodesmic reactions 193
    3.7.7 Ionization potentials, proton affinities and electron affinities 197
    3.7.8 Density functional theory (DFT) 199
    3.7.9 Complete basis set extrapolations (CBS) 201
    3.7.10 Bond dissociation energies 202

Bibliography 207

Index 219

Chapter 1

Hydrogen Thermochemistry

Historically, experimental heats of hydrogenation have had an importance far out of proportion to the small number of research groups engaged in their measurement. Much of our quantitative understanding of such concepts as resonance, conjugation, aromaticity and antiaromaticity, molecular strain, and the more subtle aspects of molecular stability arose from hydrogenation experiments. For examples from the work of the six principal research groups in the field, see Conant and Kistiakowsky (1937), Williams (1942), Skinner (1959), Turner *et al.* (1973), Roth *et al.* (1991), and Rogers *et al.* (1998).

1.1 Definitions

When hydrogen is added across a carbon-carbon double or triple bond

$$\ce{>C=C< + H_2 -> -C(H)-C(H)-}$$

a measurable amount of heat is given off to the surroundings owing to a decrease in energy and enthalpy of the system. For a reaction carried out at constant pressure, the heat out is equal and opposite to the enthalpy change of the system ΔH. From the vast field of hydrogenation chemistry, we shall concentrate almost exclusively on the thermochemistry of simple hydrogen addition to unsaturated hydrocarbons in this work. Hydrogenation of unsaturated hydrocarbons always gives off heat, hence the molar enthalpy change of hydrogenation $\Delta_{hyd} H_{298}$ is always negative.

The molar enthalpy of formation $\Delta_f H$ of any substance is the change in enthalpy that takes place when elements combine to form one mol of that substance by a formation reaction, real or hypothetical

$$\text{Elements} \rightarrow \text{Compound}$$

In the case of hydrocarbons, the formation reaction is

$$mC + \frac{n}{2}H_2 \rightarrow C_m H_n. \tag{1.1}$$

It is possible to determine the energy change of any well defined chemical reaction $\Delta_r U$ carried out at constant volume by measuring the heat given off to or absorbed from the surroundings by suitable calorimetric means. (Heat absorbed increases the energy of the system making $\Delta_r U$ positive.) If a substance is burned at constant volume V, one speaks of its "heat" of combustion $\Delta_c U$ and if a hydrogenation is run at constant pressure P, one speaks of its heat of hydrogenation, meaning its $\Delta_{hyd} H_{298}$. Energy and enthalpy are related by the well-known definition $H \equiv U + PV$.

If the reactants and products of a formation reaction are in their standard states, the enthalpy change is a *standard* enthalpy of formation $\Delta_f H^{\circ}_{298}$

$$mC(gr) + \frac{n}{2}H_2(g, p = 0.1\,MPa) \rightarrow C_m H_n(\text{stable form}) \tag{1.2}$$

where (gr) designates the graphitic form of carbon and the stable form of the hydrocarbon is a gas (g), liquid (l) or a solid (s) according to the temperature. The unit of pressure is the pascal $Pa = Nm^{-2}$, where 10^5 Pa = 1 bar ≅ 1 atmosphere. If the reaction is carried out at or corrected to a standard temperature of 298 K (strictly defined as 298.15 K) the symbol $\Delta_f H^{\circ}_{298}$ is used. See Klotz and Rosenberg (2000), especially Table 4.1 for a more detailed discussion of standard states. An enthalpy change measured at or corrected to 298 K but not under strict standard state conditions is denoted ΔH_{298}. Each enthalpy change, $\Delta_f H^{\circ}_{298}$ and $\Delta_{hyd} H_{298}$, is a special case of the less restricted ΔH_{298}.

A reliable $\Delta_{hyd} H_{298}$ and a reliable enthalpy of formation $\Delta_f H_{298}$ of either the reactant or the product of hydrogenation leads to $\Delta_f H_{298}$ of the other participant in the reaction because $\Delta_f H°(H_2) = 0$ at 298 K and 10^5 Pa by definition.

$$\Delta_{hyd} H_{298} = \Delta_f H_{298}(\text{product}) - \Delta_f H_{298}(\text{reactant}) - \underbrace{\Delta_f H°_{298}(H_2)}_{0}. \quad (1.3)$$

Frequently, the $\Delta_f H_{298}$ of an alkane is accurately known but $\Delta_f H_{298}$ values of one or more of the several alkenes or alkynes that can be hydrogenated to it are not known. The hydrogenation experiment then yields $\Delta_f H_{298}$(alkene or alkyne) by Eq. (1.3). A simple example is that of the isomeric linear hexenes, hexadienes, and 1,3,5-hexatriene, all of which can be hydrogenated to give n-hexane. Conditions are usually such that $\Delta_f H$ calculated from hydrogenation data is very close to the enthalpy of formation of the alkene in the standard state $\Delta_f H°_{298}$. A less usual situation is determination of the $\Delta_f H_{298}$ of the product alkane by combining $\Delta_{hyd} H_{298}$ with the known $\Delta_f H_{298}$ of an alkene.

Partial heats of hydrogenation that are not directly measurable can be obtained from total heats of hydrogenation by an indirect calculation.

$$\text{1,3-butadiene} + H_2 \longrightarrow \text{1-butene} \quad (1.4)$$

The hydrogenation of 1,3-butadiene to 1-butene cannot be carried out quantitatively but we know the total heats of hydrogenation at 355 K of both components to give n-butane as the product (Conant and Kistiakowsky, 1937).

$$+ 2 H_2 \xrightarrow{-57.1} \quad (1.5a)$$

$$+ H_2 \xrightarrow{-30.3} \quad (1.5b)$$

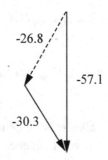

Fig. 1.1 A thermochemical cycle of hydrogenation reactions.

These results (both in kcal mol^{-1}, Chap. 2) permit one to construct a thermochemical cycle with one unknown $\Delta_{hyd}H$, shown in Fig. 1.1 as a dashed arrow. By the first law of thermodynamics, the closed path around the cycle has $\Delta H = 0$, so

$$\Delta_{hyd}H_{355}(1,3\text{-butadiene}) = -57.1 - (-30.3) = -26.8 \text{ kcal mol}^{-1}. \quad (1.6)$$

It is not always necessary to have an accurate $\Delta_f H_{298}$ for the product alkane. For example, if the $\Delta_f H_{298}$ (alkane) in Fig. 1.1 were unknown or thought to be unreliable, calculation of $\Delta_f H_{298}$ (1,3-butadiene) and $\Delta_f H_{298}$ (1-butene) would be suspect but $\Delta_{hyd}H_{298}$ of the difference between them, which is the partial reaction of 1,3-butadiene to 1-butene, would still be valid. In this way Kistiakowsky (Conant and Kistiakowsky, 1937) found the enthalpy of isomerization of *cis*- to *trans*-2-butene to be negative (*trans* more stable than *cis*) and to have $\Delta_{isom}H_{355} = -0.95 \pm 0.14$ kcal mol^{-1}. See also Turner and Garner (1957) for an analogous determination of the enthalpy difference between *exo*- and *endo*-cyclic double bonds.

No one has ever claimed that hydrogen thermochemistry will replace combustion thermochemistry but, as Kistiakowsky pointed out in his first paper on the subject (Kistiakowsky et al., 1935), reaction thermochemistry, despite its lack of generality, is a powerful tool because it can be expected to yield more accurate enthalpies of formation than combustion thermochemistry for some molecules under some circumstances. Reaction thermochemistry, especially hydrogen thermochemistry, is particularly well suited for determination of small energy differences between large molecules.

Hydrogenation thermochemistry has been effective in quantitative evaluation of quantum mechanical effects and molecular strain, for example, resonance stabilization of aromatic compounds, conjugative stabilization of linear polyenes and polyynes, and steric effects in *cis-* and *trans-* isomers. Kistiakowsky's measurement of $\Delta_{hyd}H$ of benzene (Conant and Kistiakowsky, 1937) gave the first experimental, quantitative value of the resonance energy of benzene, long known qualitatively through its reluctance to add bromine across the presumed double bonds of its Kekule structure, and given a quantum mechanical rationale by Huckel in 1931 and by Pauling in 1933. The first quantitative measure of conjugation, hyperconjugation, and strain in organic molecules arose from Kistiakowsky's determination of the differences in $\Delta_{hyd}H$ among alkenes and between monoalkenes and dienes (Conant and Kistiakowsky, 1937).

Recent advances in computational chemistry have made it possible to calculate enthalpies of formation from quantum mechanical first principles for rather large unsaturated molecules, some of which are outside the practical range of combustion thermochemistry. Quantum mechanical calculations of molecular thermochemical properties are, of necessity, approximate. Composite quantum mechanical procedures may employ approximations at each of several computational steps and may have an empirical factor to correct for the cumulative error. Approximate methods are useful only insofar as the error due to the various approximations is known within narrow limits. Error due to approximation is determined by comparison with a "known" value, but the question of the accuracy of the "known" value immediately arises because the uncertainty of the comparison is determined by the combined uncertainty of the approximate quantum mechanical result and the standard to which it is compared.

Once the validity of a quantum mechanical procedure has been established by its ability to reproduce various *accurate* experimental results, the way is clear to calculate unknown thermochemical values of unstable or explosive compounds, unsuited to classical thermochemical methods, or to calculate thermochemical properties of molecules, radicals, or ions of fleeting existence (e.g., Zavitsas, Matsunaga, and

Rogers, 2005). Herein lies a major advantage of the accuracy inherent hydrogen thermochemical results and a reason for renewed interest in the diverse but scattered literature devoted to hydrogen thermochemistry. Parts 1 and 2 of this work are devoted to experimental hydrogen thermochemistry while part 3 treats the emerging field of computational hydrogenation thermochemistry.

1.2 History

In 1935, Kistiakowsky's group (Kistiakowsky *et al.*, 1935) published the first of a series of papers on the experimental determination of $\Delta_{hyd}H$ of 51 hydrocarbons containing from 2 to 10 carbon atoms. They pointed out that, while experimental measurement of the energy of combustion $\Delta_c U$ is generally more *precise* than $\Delta_{hyd}H$ measurement, arriving at the desired enthalpy of formation $\Delta_f H_{298}$, through $\Delta_{hyd}H_{298}$ may be more *accurate* because $\Delta_{hyd}H_{298}$ is much the smaller number. Thus $\Delta_c U$ values increase monotonically with the number of carbons and the molecular weight but $\Delta_{hyd}H_{298}$ for monoalkenes, for example, do not.

At the time of Kistiakowsky's first paper, many extant thermochemical results had been measured in the nineteenth century and are now of archival interest only. Using these data, Kistiakowsky observed that it was impossible to calculate the magnitude or, in some cases, even the sign of energy differences resulting from small changes in molecular structure. This motivated an investigation into hydrogen thermochemistry by which the Kistiakowsky group determined $\Delta_{hyd}H_{298}$ at a level of precision of about ±0.1 kcal mol^{-1}. For example, they found $\Delta_{hyd}H_{298}$ for *cis*- and *trans*-2-butene to be −28.33±0.10 and −27.38±0.10 kcal mol^{-1} respectively, leading to the isomerization enthalpy $\Delta_{isom}H_{298} = -0.95\pm0.14$ kcal mol^{-1} and showing that the *trans* isomer is the more stable of the two. This was confirmed in 1951 by accurate combustion thermochemistry. Prosen, Maron *et al.* found $\Delta_c H = -647.65\pm0.29$ kcal mol^{-1} and $\Delta_c H = -646.90\pm0.23$ kcal mol^{-1} respectively, for the two isomers leading to $\Delta_{isom}H_{298}^{\circ} = -0.75\pm0.37$ kcal mol^{-1}, the difference between them. Even in this simple case, the combustion results must be 28 times as precise as the hydrogenation results to achieve the same precision in the enthalpy difference. A similar calculation for C_8 monoalkenes would require a precision ratio of 56:1, C_{12} requires a ratio of 84:1 and so on.

The series of papers written by Kistiakowsky in the late 1930s is arguably the most influential group of research papers ever published on the relationship between experimental thermochemistry and molecular structure, for it contains the first quantitative assessment of the energetic influence conferred upon molecules by the structural features conjugation, hyperconjugation, aromatic resonance, and molecular strain. Later, antiaromaticity and homoaromaticity were added to this list. These structure-energy relationships are referred to directly or indirectly in most elementary organic chemistry textbooks to this day.

Kistiakowsky's work was cut short by World War II, to be taken up briefly by Williams (1942), who carried out the first accurate $\Delta_{hyd}H_{298}$ measurements made *in solution*. Though the rationale of hydrogen thermochemistry was investigation of relatively large molecules, the gas-phase method used by Kistiakowsky had pretty nearly reached its limit in size because larger molecules are not volatile enough to be hydrogenated under the conditions he used. The change-over from gas-phase thermochemistry to solution thermochemistry was inevitable if hydrogen thermochemistry was to move in the natural direction, that is, toward the study of larger molecules and molecules less volatile than hydrocarbons. Solution hydrogenation does not suffer restrictions due to volatility and can be carried out, in principle, at any temperature. As an added complication, however, solvent interactions with both reactant and product may influence the measured $\Delta_{hyd}H_{298}$. These influences were treated in detail by Williams (1942).

Following Williams' work, Skinner's group published 3 papers in the late 1950's (Skinner, 1957). Skinner's work is noteworthy for its exploration of intramolecular energetic interactions of alkynes.

Starting contemporaneously with Skinner, Turner's group (Turner et al., 1957) published a much larger body of work ending in 1973. Turner's work produced a substantial amount of data on the relative stability of isomers, for example, *exo-endo* isomers like methylcyclopentene and methylenecyclopentane.

More recent work in hydrogen thermochemistry has been by the groups of Roth (1983) and of Rogers (Bretschneider and Rogers, 1970). Typically, work of the Roth group is characterized by numerous measurements on structurally interesting molecules, high accuracy (typically ±0.5 kJ = ±0.1 kcal mol^{-1}), and careful correction for small residual solvent effects.

The method we presently use (Caldwell *et al.*, 1997, Rogers *et al.*, 1998) is not as accurate as the best results obtained by the methods above, but it is very sensitive. Within accuracy limits that are never better than ±0.8 kJ mol^{-1} = ±0.2 kcal mol^{-1} (Rogers and Crooks, 1983) and may be as high as ±4 kJ mol^{-1}, we can, in favorable cases, carry out a complete series of $\Delta_{hyd}H_{298}$ determinations on a drop of liquid hydrocarbon. Because the molecules one is most interested in are often unstable, difficult to prepare, and difficult to purify (e.g., Rogers *et al.*, 1978), this can be a significant advantage.

1.3 Theory and Methodology

Kistiakowsky's apparatus was a flow calorimeter. A mixture of gaseous alkene or alkyne and hydrogen flowed into a reaction chamber filled with finely divided copper catalyst. The reaction product plus heat flowed out of the chamber through a helical glass tube connected to a suitable collection device. The purpose of the helical tube was to convey the heat of reaction to the calorimeter fluid in which the entire reaction system was immersed. Upon reaching a steady state, the amount of heat produced per unit time was measured by the rise in temperature of the calorimeter fluid. The amount of product alkane transferred to the collection device per unit time was also measured. This gave the heat of reaction per mol of reaction product collected. Calibration was by means of a low-voltage electrical heating coil.

Aside from its originality, Kistiakowsky's work is especially impressive in his anticipation of error sources and experimental problems and in the ingenious methods he used to circumvent them or to demonstrate that they had a negligible effect on the final result. In their first paper, the Kistiakowsky group reported results for over 30 full hydrogenation experiments on ethylene. Their final average was $\Delta_{hyd}H_{355}^{g} = -32.824 \pm 0.050$ kcal mol^{-1}, where the notation designates reactants and products in the gaseous state at 355 K, reported by Kistiakowsky as 82°C. Kistiakowsky also calculated $\Delta_{hyd}H_{0}^{g} =$

-31.000 ± 0.150, $\Delta_{hyd}H^g_{273} = -32.460 \pm 0.050$, and $\Delta_{hyd}H^g_{298} = -32.575 \pm 0.050 \pm 0.050$.

Almost all of Kistiakowsky's further studies were carried out at 355 K and some concern has been expressed in the literature that the difference between $\Delta_{hyd}H_{355}$ and $\Delta_{hyd}H_{298}$ might invalidate or at least diminish the usefulness of his results. This difference appears to be quite small, however. The change in constant pressure heat capacity, ΔC_P, for a reacting system

$$\text{alkene} + H_2 \rightarrow \text{alkane} \qquad (1.7)$$

consists of the difference in C_P for the alkane and that of the alkene plus $C_P(H_2)$. The C_P difference between two hydrocarbons is generally quite small, that of the alkene being slightly less than that of the alkane. The difference is partly made up by $C_P(H_2)$, so the change in heat capacity brought about by going from the reactant *system* to the product is small and the change in enthalpy of reaction with temperature is correspondingly small. The $\Delta C_P \Delta T$ contribution to $\Delta_{hyd}H_{298}$ over the temperature range specified is probably not larger than 0.25 kcal mol^{-1} (1.0 kJ mol^{-1}) per mole of hydrogen consumed in reaction 1.7. This upper limit is calculable from the ethylene data above, because the difference in C_P is larger for the ethylene-ethane pair than it is for the other hydrocarbon pairs studied by the Kistiakowsky group.

Williams (1942) gave details of the design of the first hydrogen calorimeter constructed specifically for measurements of $\Delta_{hyd}H_{302}$ in solution. It was a fairly conventional reaction calorimeter in which the reaction took place in a Dewar flask, but it was made more complicated by the necessity of maintaining the entire system under hydrogen-tight conditions. The sample was contained in a glass ampoule broken by an externally controlled mechanical device to initiate the reaction. The temperature was 302 K, negligibly different as far as this work is concerned from 298 K. Difficulties with different choices of solvent and catalyst were discussed and eventually glacial acetic acid and Pd were selected. We shall discuss solvent effects in Sec. 1.3.1.

The design used by Skinner's group entailed a reaction vessel agitated by vertical oscillation within a conventional Dewar flask. Reaction times were 10-20 minutes as contrasted to several hours reaction time in the Williams calorimeter. The solvent was glacial acetic acid and the catalyst was reduced PtO. Oscillatory calorimeters have not found favor since this work.

Turner's calorimeter was a modification of the Williams design using rotary stirring and catalyst introduction into the reaction mixture by breaking an ampoule. Turner notes a discrepancy between his results and those of Kistiakowsky for the test compound 1-heptene but does not mention the solvent effect. Turner also points out that the discrepancy is larger for polyenes than it is for monoenes, becoming quite substantial, *viz* 2.3 kcal mol^{-1} in the case of cycloheptatriene. Turner's group produced a large amount of excellent thermochemistry but it is regrettable that the solvent corrections necessary to convert their results to the gas phase (see below) were not quantitatively accessed. Careful measurements of solvent effects are a salient feature of Roth's hydrogenation studies described below.

We noticed early in our study of hydrogenation thermochemistry (Rogers and McLafferty, 1971) that glacial acetic acid, the solvent of choice at that time, is not necessary to achieve a useful reaction rate for small samples in the presence of large amounts of activated catalyst. We carried out our work using hydrocarbon solvents as the calorimeter fluid. Hydrocarbon solvents show much reduced solvent effects and have the added advantage of lower heat capacity than acetic acid, thus amplifying the thermal signal ΔT. The sample, in the form of a dilute solution in the same hydrocarbon as that chosen as the calorimeter fluid, was injected by means of a microsyringe. Very small samples and large amounts of activated catalyst in the reaction slurry reduced reaction times by 100 fold from the previous minimum to about 10–20 s. With such short reaction times, we chose to operate our microcalorimeter in isoperibol mode, that is, we made no attempt to maintain strictly constant temperature except that conditions were manipulated such that the temperature drift before and after reaction was not steep. Injection of a sample caused a typical sigmoidal temperature *vs*. time curve from which the temperature rise for the reaction was interpolated in the usual way (Rogers and Sasiela, 1973). In later versions of the calorimeter, these curves were stored in microcomputer memory and interpolated digitally (Fang and Rogers, 1992). Calibration of the thermal rise and conversion to conventional enthalpy units was by comparison to the thermal rise of a "known" such as 1-hexene or cyclohexene.

In this method, corrections are not made for the transfer enthalpy from solution to the gaseous state on the ground that solute molecules in very dilute solution with a noninteracting solvent are essentially independent of one another and mimic the ideal gas state. This assumption has been questioned but the results are confirmed by meticulous combustion experiments where they exist (Steele et al., 2002) and by high level computational values (Rogers, et al., 2004).

Roth (1980) developed a hydrogenation method which also uses alkane solvents, thus minimizing solvent interaction. His calorimeter is a very accurate commercial isothermal titration calorimeter (Christensen et al., 1973) and in his initial paper, he compared results with literature values for five olefins. The reaction vessel was maintained at constant temperature before, during, and after addition of a titer of dilute olefin solution by means of a motor-driven precision microburet. Constant temperature was maintained by means of a constant rate Peltier cooler operating in opposition to a variable heater. The heater delivered a number of fixed energy pulses to the reaction vessel controlled by a circuit with a thermistor inside the vessel. During the titration, heat generated or absorbed within the vessel disturbed the balance between the heating and cooling devices and the number of pulses delivered to the heater changed in response to the control circuit. At the beginning of the titration, a sharp change in the number of pulses was recorded and at the end, a change in the opposite direction was recorded. The size of the step function was proportional to the enthalpy change within the reaction chamber. This was converted to conventional enthalpy units by calibration against the heat output from a "reference" resistance heater. An advantage of this ingenious arrangement is that optimum precision could be obtained by adjusting either the voltage to the reference heater or the speed of the microburet so that the size of the hydrogenation or solution plateau (or depression) was equal to that of the reference plateau.

In his initial paper, Roth used acetic acid, methanol, and cyclohexane as solvents and Pd over carbon support, Pt/C, Pt, Pd/BaSO$_4$, and PtO$_2$. He settled on Pd/C and cyclohexane as optimal. Precision and agreement with Kistiakowsky's results were in the region of 0.1–0.2 kcal mol^{-1}.

1.3.1 *Correction to the gaseous state*

Most contemporary molecular mechanical or quantum mechanical calculations give the properties of a single molecule isolated from all others, that is, a molecule in the ideal gaseous state. Over the years, enthalpies of hydrogenation have been measured by several experimental strategies using a variety of solvents. Only Kistiakowsky's early work on relatively small molecules involved hydrogenation in the gaseous phase. For reasons of limited vapor pressure of substances of current interest, his methods are unlikely to be used again.

Given that some sort of solution thermochemistry is necessary, one must either make the corrections outlined by Williams (1942) and Fuchs and Peacock (1979) or make use of uncorrected or partially corrected data with an awareness of the enthalpic corrections that have been ignored.

Fuchs and Peacock (1979) give the thermodynamic cycle

$$
\begin{array}{ccc}
\text{alkene(g)} & \xrightarrow{\Delta_{hyd}H(g)} & \text{alkane(g)} \\
\downarrow \Delta H(v \to S) & & \downarrow -\Delta H(v \to S) \\
\text{alkene(soln)} & \xrightarrow{\Delta_{hyd}H} & \text{alkene(soln)}
\end{array}
$$

where (g) denotes the gas phase, $\Delta_{hyd}H$ is the uncorrected enthalpy of hydrogenation measured in solution, and $\Delta H(v \to S)$ is the enthalpy of transfer from vapor to solvent

$$\Delta H(v \to S) = \Delta_{soln}H - \Delta_{vap}H. \qquad (1.8)$$

The influence of experimental strategy on the solvent correction is seen by contrasting the experiments of Williams with those of Turner's group. Williams gives the equation

$$\Delta H_g = \Delta H - (_X L_v - _Y L_v) - L_s \qquad (1.9)$$

where L designates the molar "latent heats" of vaporization and solution respectively, subscripted X denotes the reactant and Y denotes the product. In the notation we use here, $L_v = \Delta_{vap}H$, $L_S = \Delta_{soln}H$, and

$$\Delta_{hyd}H_{298}(g) = \Delta_{hyd}H - \Delta_{vap}H_{298}(\text{alkene}) + \Delta_{vap}H_{298}(\text{alkane})$$
$$-\Delta_{soln}H_{298}(\text{alkane}) \quad (1.10)$$

for the reaction of a liquid alkene to the corresponding liquid alkane where the temperature is taken to be 298 K. If the reactant and product are both solids, Williams's Eq. 2 is, in our notation,

$$\Delta_{hyd}H_{298}(g) = \Delta_{hyd}H - \Delta_{fus}H_{298}(\text{alkene}) + \Delta_{fus}H_{298}(\text{alkane})$$
$$-\Delta_{vap}H_{298}(\text{alkene}) + \Delta_{vap}H_{298}(\text{alkane}) - \Delta_{soln}H_{298}(\text{alkane}) \quad (1.11)$$

where $\Delta_{fus}H_{298}$ denotes the enthalpy of fusion of the solid at 298 K.

As a trial case, Williams chose hydrogenation of 1-heptene. He equilibrated Pt or Pd catalyst in glacial acetic acid under dry H_2 at a pressure just over ambient. Upon breaking an ampoule containing 1 to 1.5 g (accurately weighed) of 1-heptene, $\Delta_{hyd}H$ was measured for the reaction

$$\text{1-heptene(l)} + H_2(g) \rightarrow \text{n-heptane (s)}$$

where (l) denotes the liquid state and (s) denotes n-heptane in solution. The temperature was 302.1 K. Results using Pt catalyst took up more than one equivalent of H_2 and were deemed unsuitable. The result obtained for 3 runs using Pd catalyst was

$$\Delta_{hyd}H = -28.280 \pm 0.127 \text{ kcal mol}^{-1}.$$

The enthalpy of solution of n-heptane in glacial acetic acid was found in a separate experiment to be 1.436 ± 0.006 kcal mol^{-1}. Williams took enthalpies of vaporization from the literature, obtained by standard classical thermodynamic means, and corrected them to a temperature of 302.1 K to coincide with his $\Delta_{hyd}H$ measurements. This led to a difference

$-\Delta_{vap}H_{302.1}(\text{1-heptene}) + \Delta_{vap}H_{302.1}(n\text{-heptane}) = -0.257 \pm 0.152 \text{ kcal mol}^{-1}$.

The corrected $\Delta_{hyd}H_{298}(g)$ is

$\Delta_{hyd}H_{298}(g) = -28.280 \pm 0.127 - 0.257 \pm 0.152 - 1.436 \pm 0.006 \text{ kcal mol}^{-1}$

which leads to $\Delta_{hyd}H_{302.1}(g) = -29.973 \pm 0.255 \text{ kcal mol}^{-1}$. This value was further corrected from 302.1 K to 355.1 K in order to coincide with Kistiakowsky's measurements. The result was

$$\Delta_{hyd}H_{355.1}(g) = -30.195 \pm 0.285 \text{ kcal mol}^{-1}.$$

This is to be compared to Kistiakowsky's measurement of

$$\Delta_{hyd}H_{355.1}(g) = -30.137 \pm 0.037 \text{ kcal mol}^{-1}.$$

Agreement is well within experimental uncertainty. The uncertainty in the difference between the two enthalpies of vaporization is largest and the enthalpies of vaporization contribute least to the result. It is to be anticipated that these enthalpies will not be known for most compounds of future interest, and the temperature at which these experiments were carried out, 302.1 K is closer to the temperature of interest, 298.1 K, so Williams simply dropped the vaporization and temperature corrections to obtain $\Delta_{hyd}H_{302.1}(g) = -29.716 \pm 0.133 \text{ kcal mol}^{-1}$. The precision of this measurement is about 0.5% or 0.15 kcal mol^{-1}. He estimates the over-all uncertainty as $\pm 1.0\% = 0.3$ kcal mol^{-1}.

Turner's experimental strategy leads to a slightly different correction pattern. In a slight variation in procedure, Turner's group broke the sample ampoule before a similar catalyst ampoule. They re-equilibrated the system before breaking the catalyst ampoule and measured the temperature change. The heat of activation of the catalyst, measured in a separate experiment, was subtracted from the total measured heat. In this way, the liquid → solution reaction of Williams was replaced by a solution → solution reaction, partly escaping the thermal effects of dissolution of the liquid or solid sample, hence partly escaping the necessity for their correction. The escape was not complete however, because, during hydrogenation in glacial acetic acid, a fairly strong solvent-alkene complex is broken up with formation of a solvent-alkane complex which is weak if it exists at all. Breaking up the complex is endothermic to the extent of about 0.7 kcal mol^{-1} per mole of H$_2$, causing the measured $\Delta_{hyd}H_{298}$ to be 0.5 kcal mol^{-1} less exothermic than it would be in the absence of a solvent effect.

1.4 Accuracy

Literature values of $\Delta_{hyd}H_{298}$ are numerous and accurate. There are about 500 known enthalpies of hydrogenation, but there has been no compilation of results since that of Jensen more than a quarter-century ago (Jensen, 1976).

Many of the $\Delta_{hyd}H_{298}$ values in the literature have an experimental uncertainty of 1 kJ mol^{-1} or less. For example, measurements of $\Delta_{hyd}H_{298}$ (1,3,5-cycloheptatriene) using three different techniques in experiments separated by 44 years (Conn et al., 1939, Turner et al., 1973, Roth et al., 1983), have an arithmetic mean experimental uncertainty and a range, when corrected for solvent effects and temperature differences to 298 K, of 1.1 kJ mol^{-1}. This level of accuracy is important now that advances in computational methods are such that we may have to decide, between two computational methods that differ by 2 kJ mol^{-1}, as to which is in better agreement with experiment (Martin, 1998, Raghavachari et al., 1997). In matters of this kind, the old standard "thermochemical accuracy" of 1 kcal mol^{-1} (4.2 kJ mol^{-1}) no longer suffices. $\Delta_{hyd}H$ values are often accurate and precise enough to meet current stringent accuracy standards for use in evaluation and comparison of computational procedures, and for parameterization of empirical or semi-empirical computational methods (e.g., Rogers et al., 1979).

1.5 Applications

A simple determination of resonance stabilization in benzene is shown in Fig. 2. Suppose we take three moles of cyclohexene and one mole of benzene as two different thermodynamic systems, each having three moles of double bonds. Upon hydrogenation, the heat output is very different for the two systems. The cyclohexene system suffers a decrease of $\Delta_{hyd}H_{298}$ = 3(-28.6) = -85.8 kcal mol^{-1} = -359 kJ mol^{-1}, while the benzene system has $\Delta_{hyd}H_{298}$ of only -49.8 kcal mol^{-1} = -208 kJ mol^{-1}. Benzene has a lower enthalpy than we might expect it to have and is said to be more stable by 36 kcal mol^{-1} = 151 kJ mol^{-1} *relative to 3 moles of cyclohexene*. Stability in benzene is conventionally ascribed to "resonance", a quantum mechanical property related to the release of spatial constraints on the electrons in benzene relative to those in cyclohexene, i.e., electron delocalization.

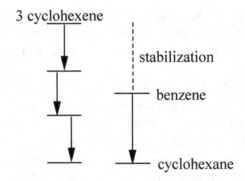

Fig. 1.2 A simple measurement of resonance stabilization. $\Delta_{hyd}H$(benzene) is about 150 kJ mol^{-1} less exothermic than 3 × $\Delta_{hyd}H$(cyclohexene).

Relative methods of calculating resonance stabilization or indeed any relative stabilization, destabilization or strain enthalpies are open to criticism, largely on the choice of a reference state (Fishtik and Datta, 2003). A different reference state, for example three ethene molecules in place of three cyclohexene molecules, gives a different "resonance stabilization" (48.6 kcal mol^{-1} = 203 kJ mol^{-1}). Thermochemical criteria have, however, the virtues of simplicity and of basing theoretical constructs on enthalpies one can, at least in principle, measure directly, as distinct from things (ring currents, *etc.*) that we suppose are correlated with enthalpy. Theoretical artifacts ("hyperconjugation", "virtual states", *etc.*) have the disadvantage that they can be molded to fit an author's preconceived notions (e.g., Jarowski *et al.*, 2004). Thermodynamic numbers are what they are, impervious to argument. Stability is, in the final analysis, a thermodynamic property.

The thrust of much modern preparative chemistry is toward synthesis and purification of milligram amounts of product. This has great advantage in making micro-purification methods feasible, often providing the experimental thermochemist with 99+ % samples of rare, often unstable compounds, albeit in very small amounts. Synthetic virtuosity feeds the interest of theoretical chemists, but it places correspondingly great demands on the sensitivity of the thermochemical methods to be employed in finding $\Delta_{hyd}H$ (or $\Delta_c H$), and ultimately the desired enthalpy of formation $\Delta_f H_{298}$.

1.6 Details of Calorimeter Construction

Given the availability of modern instrumentation, the contemporary chemist is spared the meticulous and laborious procedures followed in early hydrogen thermochemistry (see, for example, Kistiakowsky, et al., 1935, 1936). Only two more recent hydrogen calorimeter implementations are recommended to the scientist who wishes to pursue this line of research, one a commercial instrument (Tronac Inc. Orem, UT, USA) and one that can be constructed from standard laboratory equipment with a few modifications.

Roth's group has achieved excellent results using a commercial, electrically calibrated, titration calorimeter (Roth, et al., 1980 and following papers). Some of the instrumental details of the calorimeter have been reviewed (Christensen, et al., 1973) along with details of its testing on standard substances (Roth and Lennartz, 1980). In this work, the measured enthalpy of *solution* of, for example, isooctane in cyclohexane becomes smaller during a titration run owing to the change in the nature of the cyclohexane-isooctane mixture as isooctane is added. This nearly linear change was extrapolated to the "first" solution enthalpy, that is, $\Delta_{soln}H$ at infinite dilution. A comparable trend in $\Delta_{hyd}H$ during the during sample addition was not observed.

Because the commercial instrument used by Roth has been fully documented (Christensen et al., 1973), the "home made" device will be more completely described here. We have constructed calorimeters from a design that has evolved with use over a number of years (Rogers, et al.,

1971 and following papers), Throughout the evolution of the design, principles of simplicity, economy, miniaturization, and, above all, safety have been followed. Simplicity, economy, and safety need no recommendation. The categories given are not mutually exclusive. For example, the smaller the calorimeter and its attendant hydrogen carrying apparatus, the less hydrogen there is to be controlled, the less hydrogen that will escape in the event of an accident, and the safer the entire procedure is. In 30 years of hydrogenation research, we have suffered no injuries.

Miniaturization is especially advantageous in an era when compounds with extraordinarily interesting structural and thermochemical properties are being synthesized but only in very small amounts. Because the first principle of all calorimetry is that the sample must be well defined and pure (or at least have a small amount of known impurity), microcalorimetry permits use of a wider range of contemporary purification techniques, especially preparative gas chromatography, than traditional calorimetry. There is, of course, no reason to suppose that the evolutionary process of hydrogen calorimeter design cannot be continued to produce smaller, safer, and possibly more accurate instruments.

The calorimeter used in our laboratory was a 25 ml Erlenmeyer flask sealed by means of a serum stopper (Z 10,076-5, Aldrich Chem. Co., P. O. Box 2060, Milwaukee, Wisconsin 53201, USA) containing about 10 ml of a stirred slurry of catalyst and a noninteracting, nonpolar solvent, typically *n*-hexane, but possibly one of many other choices, as the occasion demanded.

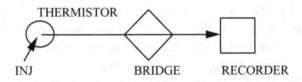

Fig. 1.3 Schematic diagram of the calorimeter.

The temperature monitoring circuit consisted of a thermistor (YSI.com) protruding through the serum stopper into the slurry and connected as one arm of a conventional Wheatstone bridge powered by an ordinary dry cell, the output of which was fed into a millivolt recorder (Rogers, 1980). Injection of sample or standard causes a rapid reaction in the flask at the left of the schematic diagram in Fig. 1.3, which brings about a change in the resistance of the thermistor and a change in the voltage output of the bridge.

The catalyst charge was 200–300 mg of 5% or 10% Pd or Pt on charcoal (Aldrich). Stirring rates of approximately 400–600 rpm were used (Multistirrer 6 or equivalent, www.velp.com). The exact stirring rate is flexible but it must be held constant during a run to achieve a moderate, steady temperature drift before and after the temperature rise due to hydrogenation. The reaction flask, wrapped in an insulating blanket, was firmly held in place by a styrofoam box or plain 600 mL beaker glued to the upper surface of the stirrer so that only the inlet septum protruded slightly from the insulation. In constructing the box, the walls should be thick but the bottom should be thin or there should be no bottom at all in order not to interfere with stirring. This, of course, requires a stirrer that produces very little heat during constant operation over the entire experiment.

Hydrogen was admitted to the reaction flask by means of a hypodermic needle thrust through the septum of the serum stopper. Upon admission of hydrogen, a sharp temperature rise of several mK occurs due to activation of the catalyst. After a few minutes, the temperature begins to drift slowly down. The calorimeter is ready for the first injections of sample and standard when the temperature drift as seen on the recorder or microcomputer monitor (see below) approximates the drift lines in Fig. 1.5. Over a time span of 10–30 s, the drift should be well approximated by a linear function.

The Tygon tube connecting the needle to the hydrogen source should be of as small bore as possible and should be as short as possible. Although we did not use it, if we were to build another hydrogen calorimeter, capillary tubing would be used throughout in order to keep the volume of hydrogen small. Lecture bottles of hydrogen are available (sigmaaldrich.com CAS#295396-56L). Under an internal operating pressure of ~1 atm over ambient pressure, the stopper sometimes pops out. This can be prevented by a drop of "Krazy Glue ®" (Elmer's Products Inc. Columbus OH 63215-3799). **Safety precaution**: The stopper must be forced very firmly into the mouth of the flask. Always wrap the flask in a towel or wear protective gloves to prevent cuts in case the flask breaks.

Samples consisting of 20–40 μL of an approximately 10% solution (depending on the degree of unsaturation) were injected through the septum using a 25 or 50 μL microsyringe fitted with a KelF adaptor (Hamilton Co. P. O. Box 10030, Reno, Nevada 89510, USA).

Determination of $\Delta_{hyd}H_{298}$ of an unknown alkene or alkyne was by comparison with that of a standard, usually $\Delta_{hyd}H_{298}$ (1-hexene) = –30.25 kcal mol^{-1} (Skinner and Snelson, 1959, Rogers, 1979). Injections of sample were made in alternation with injections of the standard in order to measure the ratio of output heats $q(\text{sample})/q(\text{standard})$. Waiting a few moments between injections was sufficient to reestablish a baseline as in Fig. 1.5. Statistical analysis shows that 9 ratios (18 injections in all) are optimum in order to take enough samples for statistical validity of the 95% confidence limit calculation but to avoid needless and redundant measurements.

The heat output of the standard, $q(\text{standard})$ was adjusted to within a few % of the heat output of the unknown sample $q(\text{sample})$ by selecting the concentration of the standard solution. Most of the time, $q(\text{sample})$ could be guessed fairly closely from that of analogous cases, enabling us to select the right concentration of the standard solution at the outset. If the first guess was wrong, results from the first few injections enabled us to calculate the right concentration for the standard. A new standard solution was made up to take the 9 ratio measurements necessary to complete the experiment.

Once $q(\text{sample}) \cong q(\text{standard})$, one already knows $\Delta_{hyd}H_{298}$ of the sample to within a few percent. The rest of the experiment is devoted to reducing this uncertainty as much as possible through the simple ratio

$$\frac{\Delta_{hyd}H_{298}(\text{sample})}{30.25} = \frac{q(\text{sample})}{q(\text{standard})}. \quad (1.12)$$

Given the calorimeter dimensions and standard and sample concentrations above, the voltage output of the bridge was 1–2 mv. Rise time of the voltage from a steady drift before reaction to a steady drift after reaction was 10–12 s. Temperature changes were not calculated; only the ratio of the heats in Eq. (1.12) was needed. Voltage output from the bridge can be input to a potentiometric recorder so as to obtain pairs of standard-sample reactions which give response curves like those in Fig. 1.5. Fitting a pair of verticals to the extrapolated linear drifts before and after reaction for a series of 18 alternating injections of standard and sample leads to a series of 9 ratios of verticals which is the series of q(sample) /q(standard) ratios we seek for 9 replicate solutions of Eq. (1.12). Calculation of the mean, standard error, and standard deviation follow routinely.

1.7 Design Modifications

We used either a commercial Wheatstone bridge or one of several that were easily assembled on a circuit board from commonly available parts.

Working from recorder output is inconvenient, and in some case subjective (as in where, exactly, to draw the vertical when you know the result you want). For this reason we substituted the A-D converter and microcomputer in Fig. 1.4 for the recorder.

The topic of computer interfacing is a large and changing one. An early and very readable introduction to the field was given in the *Bugbooks* by Larsen (1975) and Rony (1976), who were active at the very beginning of the microcomputer revolution. More recently, An (1998) has reviewed the subject.

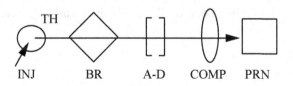

Fig. 1.4 Schematic diagram of the calorimeter with an interfaced computer.

Typically, the bridge output was conveyed through an A-D converter to the computer COMP, where several hundred points on the curve were stored digitally and the curve was displayed on the computer monitor. Operational amplifiers were also used between the bridge and A-D converter with satisfactory results. Depending on the converter properties, the amplifier may or may not be necessary.

The computer was programmed to perform a linear least squares fit to the temperature drift before and after reaction to give the equation of the upper and lower straight lines in Fig. 1.5. The difference between the two linear least squares functions at the vertical passing through the point 2/3 up the reaction thermogram (Rogers and Dejroongruang, 1988) is proportional to the heat of reaction.

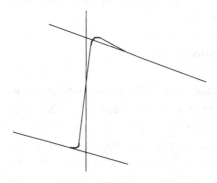

Fig. 1.5 Thermogram with least squares extrapolations.

Alternating sample and standard injections, having previously input the $\Delta_{hyd}H_{298}$ of the standard to the computer, the computer carried out a simple arithmetic program to calculate q(sample)/q(standard) and hence $\Delta_{hyd}H_{298}$ by Eq. (1.12). The result was stored digitally, and printed out to PRN to obtain a hard copy record.

Chapter 2

Experimental Results

2.1 Enthalpies of Hydrogenation

Experimental uncertainties are in parentheses. Some common names are given. In the annotated text, the symbol $\Delta_{hyd}H$ should be understood to be a negative quantity and to refer to 298 K, unless otherwise stated.

Table 2.1 Enthalpies of hydrogenation.

| Compound | $\Delta_{hyd}H$ | |
Temperature/K	kcal mol^{-1}	kJ mol^{-1}
C_2H_2		
(1) Ethyne		
Acetylene		
355	75.1(0.2)	314.0(0.6)
298	74.6(0.2)	312.1(0.6)

Measurements of $\Delta_{hyd}H$ by Kistiakowsky's group were made using a flow calorimeter. Reactants were in the gaseous state. The temperature was usually, but not always, 355 K.

Conn, J. B.; Kistiakowsky, G. B.; Smith, E. A. *J. Amer. Chem. Soc.* **1939**, *61*, 1868–1876. Temperature correction to 298 K was made by Cox, J. D.; Pilcher, G., 1970. *Thermochemistry of Organic and Organometallic Compounds.* Academic Press, New York.

C_2H_4

(1) Ethene
    Ethylene
355 K	32.82(0.05)	137.3(0.2)
298 K	32.58(0.05)	136.3(0.2)
273 K	32.46(0.05)	135.8(0.2)
0 K	31.00(0.15)	129.7(0.6)

The actual experimental measurement at 355 K was used along with C_P data to obtain the values at temperatures other than 355 K. The authors expressed some reservations with regard to the accuracy of the C_P data, hence the uncertainty limits are greater at 0 K than at 355 K.

Kistiakowsky, G. B.; Romeyn, Jr., J. R.; Smith, H. A.; Vaughan, W. E. *J. Amer. Chem. Soc.* **1935**, *57*, 65–75.

(1') Ethylene
298 K	32.60(0.05)	136.3(0.2)

Results are by indirect calculation from an equilibrium study. The motivation for this study was cross-checking the hydrogenation results, which are verified by this result.

Kistiakowsky, G. B.; Nickle, A. G. *Disc. Faraday Soc.* **1951**, 175–187.

C_3H_4

(1) Propyne
    Methyl acetylene
298 K	69.7(0.2)	291.6(0.6)

Conn, J. B.; Kistiakowsky, G. B.; Smith, E. A. *J. Amer. Chem. Soc.* **1939**, *61*, 1868–876. The temperature correction to 298 K is probably about 0.4 kcal mol^{-1}.

(2) Propadiene
    Allene
355 K	71.28(0.2)	298.2(0.8)

The authors cite evidence of slight polymerization. It is interesting to compare this result with that for propyne. One anticipates isomerization equilibriums between allenes and alkynes in general.

Kistiakowsky, G. B.; Ruhoff, J. R.; Smith, H. A.; Vaughan, W. E. *J. Amer. Chem. Soc.* **1936**, *58*, 146–153.

C_3H_6

(1)   Propene
      355 K   30.12(0.1)   126.0(0.1)

Hydrogenation was carried out in the gaseous phase.

Kistiakowsky, G. B.; Ruhoff, J. R.; Smith, H. A.; Vaughan, W. E. *J. Amer. Chem. Soc.* **1935**, *57*, 876–882.

(1') Propene
      298 K   29.85(0.05)   124.9(0.2)

Results are by indirect calculation from an equilibrium study. The motivation for this study was cross-checking the hydrogenation results. The results are very slightly outside of combined experimental error.

Kistiakowsky, G. B.; Nickle, A. G. *Disc. Faraday Soc.* **1951**, 175–187.

C_4H_4

(1)   1-Butene-3-yne
      298 K   100.8(0.5)   421.7(2.1)

Roth, W. R.; Adamczak, O.; Breuckman, R.; Lennartz, H.-W.; Boese, R. *Chem. Ber.* **1991**, *124*, 2499–2521.

C_4H_6

(1)   2-Butyne
      355 K   65.6 (0.3)   272.4(1.3)

Conn, J. B.; Kistiakowsky, G. B.; Smith, E. A. *J. Amer. Chem. Soc.* **1939**, *61*, 1868–1876.

(2)   1,3-Butadiene
      355 K   57.07(0.1)   238.8(0.4)

The difference between $\Delta_{hyd}H$ (1,3-butadiene) and twice $\Delta_{hyd}H$ (1-butene), $2(-30.34) - (-57.07) = -3.6$ kcal mol^{-1} = -15.1 kJ mol^{-1} is often called the "conjugation stabilization energy" and is ascribed to delocalization of electron probability density in the diene.

Kistiakowsky, G. B.; Ruhoff, J. R.; Smith, H. A.; Vaughan, W. E. *J. Amer. Chem. Soc.* **1936**, *58*, 146–153.

C_4H_8

(1)   1-Butene
      355 K   30.34(0.1)   126.9(0.4)

The temperature correction from 355 K to 298 K is estimated as 0.25 kcal mol^{-1} in this paper.

Kistiakowsky, G. B.; Ruhoff, J. R.; Smith, H. A.; Vaughan, W. E. *J. Amer. Chem. Soc.* **1935**, *57*, 876–882.

(2)   2-Butene, (E)
      355 K   27.62(0.1)   115.6(0.4)

The internal double bond in 2-butene is stabilized by 2.7 kcal mol^{-1} = 11.3 kJ mol^{-1} relative to the terminal double bond in 1-butene.

Kistiakowsky, G. B.; Ruhoff, J. R.; Smith, H. A.; Vaughan, W. E. *J. Amer. Chem. Soc.* **1935**, *57*, 876–882.

(3)   2-Butene, (Z)
      cis-2-Butene
      355 K   28.57(0.1)   119.5(0.1)

The difference in $\Delta_{hyd}H$ between the E and Z isomers of 2-butene is the archetypal E-Z isomerization energy, 1.0 kcal mol^{-1} = 4.2 kJ mol^{-1}. The E isomer is more stable than the Z isomer, presumably due to methyl crowding in the latter.

Kistiakowsky, G. B.; Ruhoff, J. R.; Smith, H. A.; Vaughan, W. E. *J. Amer. Chem. Soc.* **1935**, *57*, 876–882.

(4)   Methylpropene
      Isobutene
      355 K   28.39(0.1)   118.8(0.3)

The difference in $\Delta_{hyd}H$ between 1-butene and methyl propene is often called a "hyperconjugation energy", in this case, 2.0 kcal mol^{-1} = 8.2 kJ mol^{-1}. In general, a methyl group alpha to a double bond is said to stabilize the reactant (See, however, e. g., Rogers, D. W. *Tetrahedron Letters* **1987**, *28*, 1967.)

Kistiakowsky, G. B.; Ruhoff, J. R.; Smith, H. A.; Vaughan, W. E. *J. Amer. Chem. Soc.* **1935**, *57*, 876–882.

C_5H_6

(1) Cyclopentadiene
     355 K   50.86(0.2)   212.8(0.8)

The difference in $\Delta_{hyd}H$ (cyclopentadiene) and twice $\Delta_{hyd}H$ (1-butene) is 9.8 kcal mol^{-1} = 41.1 kJ mol^{-1}. The enthalpy of the reactant state is lowered by conjugation and that of the product state is raised by hydrogen atom crowding in the 5-membered ring. Both factors narrow the gap in enthalpy between reactant and product, which is $\Delta_{hyd}H$.

Kistiakowsky, G. B.; Ruhoff, J. R.; Smith, H. A.; Vaughan, W. E. *J. Amer. Chem. Soc.* **1936**, *58*, 146–153.

(1') Cyclopentadiene
      298 K   51.4(0.1)   215.1(0.4)

Roth, W. R.; Klaerner F.-G.; Lennartz, H.-W. *Chem. Ber.* **1980**, *113*, 1818–1829.

(2) 3-Penten-1-yne, (E)
     trans-pent-3-en-1-yne
     298 K   96.0(0.4)   401.7(1.7)

$\Delta_{hyd}H$ was measured in glacial acetic acid and corrected by a separate heat of solution measurement. The value given is for the liquid → liquid reaction. Experimental uncertainty is large for alkynes and ene-ynes because experimental error increases linearly with $\Delta_{hyd}H$. The ene-yne conjugation stabilization relative to propyne and 2-butene is small, $-96.0 - (-97.3) = 1.3$ kcal mol^{-1} = 5.4 kJ mol^{-1}, that is, $\Delta_{hyd}H$ is smaller in magnitude by this amount than it would be for a hypothetical molecule with no conjugation interaction.

Skinner, H. A.; Snelson A. *Trans. Faraday Soc.* **1959**, *55*, 404–407.

(3) 3-Penten-1-yne, (Z)
     cis-Pent-3-en-1-yne
     298 K   95.60(1.1)   400.0(4.6)

$\Delta_{hyd}H$ was measured in ethanol and corrected by a separate heat of solution measurement. The value given is for the liquid → liquid reaction. It is noteworthy that the E-Z stability ratio is opposite to that of the 2-butenes, suggesting that Z alkyne groups are less bulky than either Z methyl groups or H atoms. The authors caution that the E-Z enthalpy difference is smaller than experimental uncertainty and that this conclusion should be regarded with some skepticism.

Skinner, H. A.; Snelson A. *Trans. Faraday Soc.* **1959**, *55*, 404–407.

(2') 3-Penten-1-yne, (E)
      298 K   97.0(0.3)   405.8(1.3)

Roth et al. comment that thermochemical conjugation stabilization between the double bond and the triple bond is very much smaller than would be expected by analogy to 1,3-butadiene. If $\Delta_{hyd}H$ of an isolated terminal triple bond is taken as about –69.5 kcal mol^{-1} and that of an internal double bond is ~ –28.0 kcal mol^{-1}, the expected $\Delta_{hyd}H$ of a *completely nonstabilized* 3-penten-1-yne, (E) would be –97.5 kcal mol^{-1}.

Roth, W. R.; Adamczak, O.; Breuckman, R.; Lennartz, H.-W.; Boese, R. *Chem. Ber.* **1991**, *124*, 2499–2521.

(3')   3-Penten-1-yne, (Z)
        cis-3-penten-1-yne
        298 K   96.8(0.1)   405.0(0.4)

The more precise values 2' and 3' confirm the earlier conclusions 2 and 3.

Roth, W. R.; Adamczak, O.; Breuckman, R.; Lennartz, H.-W.; Boese, R. *Chem. Ber.* **1991**, *124*, 2499–2521.

(4)   Bicyclo[2.1.0]pent-2-ene
        298 K   98.0(0.5)   410.0(2.1)

Difficulties with isomerization of bicyclo[2.1.0]pent-2-ene were noted and overcome by the authors who also note that antiaromatic destabilization of bicyclo[2.1.0]pent-2-ene is manifested both in its thermochemistry and in its augmented dienophilic reactivity relative to cyclopentadiene. They refer to bicyclo[2.1.0]pent-2-ene as "pseudobicyclobutadiene" because a cyclopropane ring replaces the double bond in cyclobutadiene. The hydrogenation product is cyclopentane.

Roth, W. R.; Klaerner F.-G.; Lennartz, H.-W. *Chem. Ber.* **1980**, *113*, 1818–1829.

C_5H_8

(1)   1,2-Pentadiene
        298 K   62.6(0.1)   261.9(0.4)

Roth, W. R.; Adamczak, O.; Breuckman, R.; Lennartz, H.-W.; Boese, R. *Chem. Ber.* **1991**, *124*, 2499–2521.

(2)   1,3-Pentadiene
        355 K   54.1(0.2)   226.4(0.6)

1,3-pentadiene has a terminal double bond and an internal double bond, which are conjugated. One can estimate its $\Delta_{hyd}H$ as $-30.3 + (-27.6) + 3.6 = -54.3$ kcal mol^{-1} = -227.2 kJ mol^{-1}, that is, for simple hydrocarbons, a rough kind of additivity is observed. Additivity rules will break down as more complicated cases are encountered later in this table.

Dolliver, M. A.; Gresham, T. L.; Kistiakowsky, G. B.; Vaughan, W. E. *J. Amer. Chem. Soc.* **1937**, *59*, 831–841.

(3) 1,4-Pentadiene
     355 K   60.79(0.2)   254.3(0.6)

$\Delta_{hyd}H$ for two terminal double bonds is within experimental uncertainty of twice the value for the double bond in 1-butene.

Kistiakowsky, G. B.; Ruhoff, J. R.; Smith, H. A.; Vaughan, W. E. *J. Amer. Chem. Soc.* **1936**, *58*, 146–153.

(4) Cyclopentene
     355 K   26.9(0.1)   112.6(0.4)

$\Delta_{hyd}H$ of cyclopentene to produce cyclopentane is smaller than it is in cyclohexene because of crowding of hydrogen atoms in the smaller product molecule relative to the larger one.

Dolliver, M. A.; Gresham, T. L.; Kistiakowsky, G. B.; Vaughan, W. E. *J. Amer. Chem. Soc.* **1937**, *59*, 831–841.

(4') Cyclopentene
      298 K   26.0(0.4)   109.0(1.8)

$\Delta_{hyd}H$ was measured in acetic acid solution. Solvent effects for alkenes in this experimental method cause the measured $\Delta_{hyd}H$ to be very roughly 0.7 kcal mol^{-1} less exothermic than gas-phase values, which would lead to a corrected value of $-26.74(0.44)$ kcal mol^{-1} = -111.9 kJ mol^{-1}. This entry is the mean of 12 separate thermochemical experiments made over a 10-year period. An extra digit is carried in the uncertainty when this measurement is used as an indicator of the overall uncertainty of the method.

Turner, R. B.; Jarrett, A. D.; Goebel, P.; Mallon, B. J. *J. Amer. Chem. Soc.* **1973**, *95*, 790–792.

(4") Cyclopentene
      298 K   26.2(0.2)   109.6(0.8)

Hydrogenation was carried out in glacial acetic acid and corrected for solvent effects. In many cases authors report their results to only three significant digits or to 0.1 kcal mol^{-1} for simple alkenes. Usually, this is an accurate indicator of reliability. When four significant digits are included, the last digit is uncertain and is useful only for averaging.

Rogers, D. W.; Mc Lafferty, F. J., *Tetrahedron* **1971**, *27*, 3765–3775.

(4''')   Cyclopentene
          298 K   26.8(0.1)   112.1(0.4)

Hydrogenation was carried out in a thermoneutral solvent (cyclohexane).

Roth, W. R.; Lennartz, H.-W. *Chem. Ber.* **1980**, *113*, 1806–1817.

(4'''')   Cyclopentene
          298 K   26.9(0.1)   112.7(0.4)

The determination was carried out in a thermoneutral solvent (*n*-hexane).

Allinger, N. L.; Dodziuk, H.; Rogers, D. W.; Naik, S. N., *Tetrahedron* **1982**, *38*, 1593–1597.

(4''''') Cyclopentene
          298 K   27.03(0.17)   113.1(0.7)

The determination was carried out in a thermoneutral solvent using a microprocessor-controlled electrical calibration circuit programmed to deliver a 4.136 J energy pulse to the calorimeter over a heating time of 5.273 s (Rogers, D. W.; Munoz-Hresko, B., *Mikrochimica Acta [Wien]* **1984** *I* 417–425. Uncertainties are 95% confidence limits for 9 replicate determinations. Results are reported in four significant figures because the point of the work is precision testing.

Rogers, D. W.; Munoz-Hresko, B., *Mikrochimica Acta [Wien]* **1984** *II*, 427–435.

(4'''''') Cyclopentene
           298 K   26.72(0.31)   111.8(1.3)

The determination was carried out in a thermoneutral solvent using a microprocessor-controlled electrical calibration circuit programmed to deliver a 4.136 J energy pulse to the calorimeter over a heating time of 5.273 s (Rogers, D. W.; Munoz-Hresko, B., *Mikrochimica Acta [Wien]* **1984** *I* 417–425. Uncertainties are 95% confidence limits for 9 replicate determinations. Results are reported in four significant figures because the point of the work is precision testing.

Rogers, D. W.; Munoz-Hresko, B., *Mikrochimica Acta [Wien]* **1984** *II*, 427–435.

(5)   Methylenecyclobutane
          298 K   29.4(0.1)   123.1(0.4)

$\Delta_{hyd}H$ was measured in acetic acid.

Turner, R. B.; Goebel, P.; Mallon, B. J.; von E. Doering, W.; Coburn Jr., J. F.; Pomerantz, M. *J. Amer. Chem. Soc.* **1968**, *90*, 4315–4322.

(6) Methylcyclobutene
     298 K   28.5(0.1)   119.2(0.4)

$\Delta_{hyd}H$ was measured in acetic acid. The exocyclic-endocyclic enthalpy of isomerization (see entry (5) above) is dependent upon ring size, which determines simple angle bending imposed on the sp^2 carbon and those adjacent to it.

Turner, R. B.; Goebel, P.; Mallon, B. J.; von E. Doering, W.; Coburn Jr., J. F.; Pomerantz, M. *J. Amer. Chem. Soc.* **1968**, *90*, 4315–4322.

(7) 1,2-Dimethylcyclopropene
     298 K   43.3(1.0)   181(4)

$\Delta_{hyd}H$ was measured in acetic acid. Mixed products were observed. Pre-isomerization to the methylene intermediate was postulated. (see also Allinger, N. L.; Dodziuk, H.; Rogers, D. W.; Naik, S. N., *Tetrahedron* **1982**, *38*, 1593–1597.)

Turner, R. B.; Goebel, P.; Mallon, B. J.; von E. Doering, W.; Coburn Jr., J. F.; Pomerantz, M. *J. Amer. Chem. Soc.* **1968**, *90*, 4315–4322.

(8) Methylmethylenecyclopropane
     298 K   38.3(0.1)   160(0.5)

$\Delta_{hyd}H$ was measured in acetic acid. Some ring opening was observed and was corrected for using $\Delta_f H$ of the acyclic product. Abnormalities in the thermogram were noted.

Turner, R. B.; Goebel, P.; Mallon, B. J.; von E. Doering, W.; Coburn Jr., J. F.; Pomerantz, M. *J. Amer. Chem. Soc.* **1968**, *90*, 4315–4322.

(9) Ethylidinecyclopropane
     298 K   37.0(0.2)   155.0(0.6)

$\Delta_{hyd}H$ was measured in acetic acid. Some ring opening was observed and was corrected for using $\Delta_f H$ of the acyclic product. Abnormalities in the thermogram were noted.

Turner, R. B.; Goebel, P.; Mallon, B. J.; von E. Doering, W.; Coburn Jr., J. F.; Pomerantz, M. *J. Amer. Chem. Soc.* **1968**, *90*, 4315–4322.

(10) Bicyclo[2.1.0]pentane
     298 K   55.1(0.4)   230.7(1.5)

$\Delta_{hyd}H$ was measured in acetic acid. The product was cyclopentane. In some cases, cyclopropane rings can be hydrogenated as though they were double bonds (see also quadricyclane, Rogers, D. W.; Choi, L. S.; Girellini, R. S.; Holmes, T. J.; Allinger, N. L. J. *Phys. Chem.* **1980**, *84*, 1810–1814).

Turner, R. B.; Goebel, P.; Mallon, B. J.; von E. Doering, W.; Coburn Jr., J. F.; Pomerantz, M. *J. Amer. Chem. Soc.* **1968**, *90*, 4315–4322.

(10')   Bicyclo[2.1.0]pentane
        298 K   56.1(0.1)   234.7(0.4)

See note in entry (10).

Roth, W. R.; Klaerner F.-G.; Lennartz, H.-W. *Chem. Ber.* **1980**, *113*, 1818–1829.

(11)   Cyclopropylethane
        Vinylcyclopropane
        298 K   65.5(0.2)   274.1(0.8)

The reaction was carried out in isooctane. The reaction product was *n*-pentane.

Roth, W. R.; Kirmse, W.; Hoffmann, W.; Lennartz, H.-W. *Chem. Ber.* **1982**, *115*, 2508–2515.

C_5H_{10}

(1)   1-Pentene
        298 K   28.5(0.3)   119.2(1.3)

Hydrogenation was carried out in *n*-pentane. A systematic error of about 1.5 kcal mol^{-1} due to (endothermic) vaporization is likely.

Rogers, D. W.; Mc Lafferty, F. J., *Tetrahedron* **1971**, *27*, 3765–3775.

(2)   1-Pentene
        298 K   30.0(0.6)   125.5(2.4)

Hydrogenation was carried out in *n*-hexane. This value differs slightly from the original paper due to a revised standard $\Delta_{hyd}H$, Rogers, D. W. *J. Phys. Chem.* **1979**, *83*, 2430 (one page).

Rogers, D. W.; Skanupong, S. *J. Phys. Chem.* **1974**, *78*, 2569–2572.

(3)   2-Pentenes, (E and Z)
        355 K   27.95(0.1)   116.9(0.4)

A mixture was used in this series of experiments because of the difficulty of separating the isomers by means available at the time the experiments were carried out. The authors estimate values of 27.63 and 28.57 for the E- and Z- 2-pentenes respectively, on the basis of their ratio in the sample. The authors also determined $\Delta_f H$ (H$_2$O) by direct combination of H$_2$ and O$_2$ to test the accuracy of the method, finding – 68.380 (0.15) kcal mol^{-1} as compared to the standard value of $\Delta_f H$ – 68.315 (0.010); (CRC Handbook of Chemistry and Physics, 1993).

Kistiakowsky, G. B.; Ruhoff, J. R.; Smith, H. A.; Vaughan, W. E. *J. Amer. Chem. Soc.* **1936**, *58*, 137–145.

(4) 2-Methylbut-1-ene
     unsym-Methylethylethylene
     355 K   28.49(0.1)   119.2(0.4)

The authors give a detailed discussion of the stabilization of a double bond by groups in the α position..

Kistiakowsky, G. B.; Ruhoff, J. R.; Smith, H. A.; Vaughan, W. E. *J. Amer. Chem. Soc.* **1936**, *58*, 137–145

(5) 3-Methylbut-1-ene
     Isopropylethylene
     355 K   30.34(0.1)   126.9(0.4)

Dolliver, M. A.; Gresham, T. L.; Kistiakowsky, G. B.; Vaughan, W. E. *J. Amer. Chem. Soc.* **1937**, *59*, 831–841.

(6) 2-Methylbut-2-ene
     Trimethylethylene
     355 K   26.92(0.1)   112.6(0.4)

Kistiakowsky, G. B.; Ruhoff, J. R.; Smith, H. A.; Vaughan, W. E. *J. Amer. Chem. Soc.* **1936**, *58*, 137–145.

C_6H_4

(1) 1,5-Hexadiyne-3-ene, (E)
     298 K   168.5(0.4)   705.0(1.7)

Roth, W. R.; Adamczak, O.; Breuckman, R.; Lennartz, H.-W.; Boese, R. *Chem. Ber.* **1991**, *124*, 2499–2521.

(2) 1,5-Hexadiyne-3-ene, (Z)
     cis-1,5-Hexadiyne-3-ene
     298 K   169.4(0.4)   708.8(1.7)

Roth, W. R.; Adamczak, O.; Breuckman, R.; Lennartz, H.-W.; Boese, R. *Chem. Ber.* **1991**, *124*, 2499–2521.

C_6H_6

(1) Benzene
355 K   49.80(0.2)   208.4(0.8)

This paper reports the first experimental measurement of the resonance energy of benzene, along with the remarkable finding that the addition of one mole of hydrogen to benzene is *endothermic*.

Kistiakowsky, G. B.; Ruhoff, J. R.; Smith, H. A.; Vaughan, W. E. *J. Amer. Chem. Soc.* **1936**, *58*, 146–153.

(2) 1,5-Hexadiyne
Dipropargyl
298 K   139.4(1.0)   583.2(4.2)

$\Delta_{hyd}H$ was measured in glacial acetic acid and corrected by a separate heat of solution measurement experiment for *n*-hexane. This value is given for the liquid → liquid reaction. Oscillatory agitation of the reaction vessel achieved complete reaction times of 10 – 15 min as contrasted to several hours for the prior Williams design which used rotary stirring.

Skinner, H. A.; Snelson A., *Trans. Faraday Soc.* **1959**, *55*, 404–407.

(3) Methylenecyclopenta-1,3-dene
Fulvene
298 K   78.9(0.1)   334.3(0.4)

Roth, W. R.; Adamczak, O.; Breuckman, R.; Lennartz, H.-W.; Boese, R. *Chem. Ber.* **1991**, *124*, 2499–2521.

(4) 1,3-Hexadiene-5-yne, (Z)
cis-1,3-Hexadiene-5-yne
298 K   124.7(0.4)   521.7(1.7)

Corrected for 0.3 kcal mol^{-1} solvent effects.

Roth, W. R.; Hopf, H.; Horn, C. *Chem. Ber.* **1994**, *127*, 1781–1795.

(5) Dimethylenecyclobutene
298 K   89.8(0.3)   375.7(1.2)

$\Delta_{hyd}H$ was measured in isooctane. A correction was made of + 0.1 kcal mol^{-1} per double bond in the reactant for the heat of solution difference between reactant and product. Corrections were not made for the difference in heats of vaporization or sublimation in this study. A mixed reaction product was obtained and corrected to the major component using values of $\Delta_f H$ calculated by molecular mechanics. This is reflected in the relatively large experimental uncertainty.

Roth, W. R.; Lennartz, H.-W.; Vogel, E.; Leiendecker, M.; Oda, M. *Chem. Ber.* **1986**, *119*, 837–843.

C_6H_8

(1) 1,3,5-Hexatriene, (E)
     298 K   79.4(0.2)   332.3(0.8)

$\Delta_{hyd}H$ was measured in glacial acetic acid. The compound has received special attention because it is "linear benzene".

Turner, R. B.; Mallon, B. J.; Tichy, M.; Doering, W. von E.; Roth. W. R.; Schroeder, G., *J. Amer. Chem. Soc.* **1973**, *95*, 8605–8609.

(1') 1,3,5-Hexatriene, (E)
      298 K   80.0(0.6)   334.7(2.5)

$\Delta_{hyd}H$ was measured in *n*-hexane solution against 1-hexene as a standard. The value given is the arithmetic mean of two experiments in which $\Delta_{hyd}H$ was determined from the $\Delta_{hyd}H$ of a mixture of the isomers and glc determination of their ratio. Uncertainties are the arithmetic mean of the uncertainties for each experiment which, in turn, are twice the standard deviation from the mean for four degrees of freedom.

Fang, W.; Rogers, D. W. *J. Org. Chem.* **1992**, *57*, 2294–2297.

(2) 1,3,5-Hexatriene, (Z)
     cis-1,3,5-Hexatriene
     298 K   80.5(0.3)   336.8(1.4)

$\Delta_{hyd}H$ was measured in glacial acetic acid.

Turner, R. B.; Mallon, B. J.; Tichy, M.; Doering, W. von E.; Roth. W. R.; Schroeder, G. *J. Amer. Chem. Soc.* **1973**, *95*, 8605–8609.

(2') 1,3,5-Hexatriene, (Z)
      cis-1,3,5-Hexatriene
      298 K   81.0(0.6)   338.9(2.5)

$\Delta_{hyd}H$ was measured in *n*-hexane solution against 1-hexene as a standard. The value given is the arithmetic mean of two experiments in which $\Delta_{hyd}H$ was determined from the $\Delta_{hyd}H$ of a mixture of the isomers and glc determination of their ratio. Uncertainties are the arithmetic mean of the uncertainties for each experiment which, in turn, are twice the standard deviation from the mean for four degrees of freedom.

Fang, W.; Rogers, D. W. *J. Org. Chem.* **1992**, *57*, 2294–2297.

(3) Hexa-2,3,4-triene
 298 K   103.2(0.3)   431.8(1.3)

Destabilization of the "double allene" relative to the conjugated triene 1,3,5-Hexatriene, (E) is about 23 kcal mol^{-1}.

Roth, W. R.; Adamczak, O.; Breuckman, R.; Lennartz, H.-W.; Boese, R. *Chem. Ber.* **1991**, *124*, 2499–2521.

(4) 1,3-Cyclohexadiene
 355 K   55.37(0.1)   231.7(0.4)

$\Delta_{hyd}H$ was measured in the gas phase.

Kistiakowsky, G. B.; Ruhoff, J. R.; Smith, H. A.; Vaughan, W. E. *J. Amer. Chem. Soc.* **1936**, *58*, 146–153.

(4') 1,3-Cyclohexadiene
 298 K   53.6(0.3)   224.4(1.2)

$\Delta_{hyd}H$ was measured in glacial acetic acid. The difference between this entry and the gas-phase value preceding it (1.63 kcal mol^{-1} = 6.8 kJ mol^{-1}) is slightly larger than expected. 1,3-Cyclohexadiene is difficult to purify.

Turner, R. B.; Mallon, B. J.; Tichy, M.; Doering, W. von E.; Roth. W. R.; Schroeder, G., *J. Amer. Chem. Soc.* **1973**, *95*, 8605–8609.

(5) 1,4-Cyclohexadiene
 298 K   53.9(0.3)   225.5(1.4)

The absence of conjugation stabilization of the 1,3 relative to the 1,4 isomer is noteworthy.

Turner, R. B.; Mallon, B. J.; Tichy, M.; Doering, W. von E.; Roth. W. R.; Schroeder, G., *J. Amer. Chem. Soc.* **1973**, *95*, 8605–8609.

(5') 1,4-Cyclohexadiene
 298 K   55.6   232.6

One measurement was made, hence no experimental uncertainty is given.

Roth, W. R.; Adamczak, O.; Breuckman, R.; Lennartz, H.-W.; Boese, R. *Chem. Ber.* **1991**, *124*, 2499–2521.

(6)   1,3-Dimethylenecyclobutane
      300 K   60.0(1.0)   251.0(4.2)

$\Delta_{hyd}H$ was measured in acetic acid.

Turner, R. B.; Goebel, P.; Mallon, B. J.; von E. Doering, W.; Coburn Jr., J. F.; Pomerantz, M. *J. Amer. Chem. Soc.* **1968**, *90*, 4315–4322.

(7)   1,3-Dimethylenecyclobutane
      298 K   58.2(0.2)   243.5(0.8)

$\Delta_{hyd}H$ was measured in cyclohexane or isooctane. A correction was made of + 0.1 kcal mol^{-1} per double bond in the reactant for the heat of solution difference between reactant and product. Corrections were not made for the difference in heats of vaporization or sublimation in this study. A mixed reaction product was obtained and corrected to the major component using values of $\Delta_f H$ calculated by molecular mechanics. This is reflected in the relatively large experimental uncertainty.

Roth, W. R.; Lennartz, H.-W.; Vogel, E.; Leiendecker, M.; Oda, M. *Chem. Ber.* **1986**, *119*, 837–843.

(8)   1-Methyl-3-methylenecyclobutene
      298 K   54.9(1.1)   229.9(4.5)

$\Delta_{hyd}H$ was measured in acetic acid. The "escape energy" of the endocyclic double bond in this compound to the exocyclic position in 1,3-dimethylenecyclobutane is about 5 kcal mol^{-1}.

Turner, R. B.; Goebel, P.; Mallon, B. J.; von E. Doering, W.; Coburn Jr., J. F.; Pomerantz, M. *J. Amer. Chem. Soc.* **1968**, *90*, 4315–4322.

(9)   Bicyclo[3.1.0]hex-2-ene
      298 K   63.1(0.2)   264.0(0.8)

The catalyst was 5% Rh/C and the reaction product was 55% methylcyclopentane and 45% cyclohexane. The thermochemical result was calculated for 100% methylcyclopentane.

Roth, W. R.; Adamczak, O.; Breuckman, R.; Lennartz, H.-W.; Boese, R. *Chem. Ber.* **1991**, *124*, 2499–2521.

(10) 3-Methylenecyclopentene

298 K 52.9(0.2) 221.3(0.8)

Roth, W. R.; Adamczak, O.; Breuckman, R.; Lennartz, H.-W.; Boese, R. *Chem. Ber.* **1991**, *124*, 2499–2521.

(11) Bicyclo[2.2.0]hex-1(4)-ene
298 K 102.1(1.9) 427.2(7.9)

The product is cyclohexane.

Roth, W. R.; Adamczak, O.; Breuckman, R.; Lennartz, H.-W.; Boese, R. *Chem. Ber.* **1991**, *124*, 2499–2521.

(12) Bicyclo[2.2.0]hex-2-ene
298 K 91.9(0.2) 384.5(0.8)

The product is cyclohexane. There is an energy difference of 10 kcal mol^{-1} between the double bond in the 1-4 position in bicyclo[2.2.0]hex-1(4)-ene and the 2-3 position in bicyclo[2.2.0]hex-2-ene.

Roth, W. R.; Klaerner F.-G.; Lennartz, H.-W. *Chem. Ber.* **1980**, *113*, 1818–1829.

(13) Bicyclo[2.1.1]hex-2-ene
298 K 40.9(0.1) 171.1(0.4)

Roth, W. R.; Adamczak, O.; Breuckman, R.; Lennartz, H.-W.; Boese, R. *Chem. Ber.* **1991**, *124*, 2499–2521.

C_6H_{10}

(1) 1-Hexyne
298 K 69.2(0.1) 289(0.4)

$\Delta_{hyd}H$ of all the linear monoalkynes in this series were measured in *n*-hexane solution.

Rogers, D. W.; Dagdagan, O. A.; Allinger, N. L. *J. Amer. Chem. Soc.* **1979**, *101*, 671–676.

(2) 2-Hexyne
298 K 65.6(0.4) 274(1.7)

$\Delta_{hyd}H$ of all the linear monoalkynes in this series were measured in *n*-hexane solution.

Rogers, D. W.; Dagdagan, O. A.; Allinger, N. L. *J. Amer. Chem. Soc.* **1979**, *101*, 671–676.

(3)   3-Hexyne
        298 K   65.1(0.3)   272.4(1.3)

$\Delta_{hyd}H$ of all the linear monoalkynes in this series were measured in *n*-hexane solution.

Rogers, D. W.; Dagdagan, O. A.; Allinger, N. L. *J. Amer. Chem. Soc.* **1979**, *101*, 671–676.

(4)   1,3-Hexadiene, (Z)
        cis-1,3-Hexadiene
        298 K   53.9(0.3)   225.5(1.3)

$\Delta_{hyd}H$ was measured in *n*-hexane solution against 1-hexene as a standard. The value given was determined from the $\Delta_{hyd}H$ of a mixture of the isomers and glc determination of the ratio of isomers in the sample.

Fang, W.; Rogers, D. W. *J. Org. Chem.* **1992**, *57*, 2294–2297.

(4')   1,3-Hexadiene, (E)
        trans-1,3-Hexadiene
        298 K   52.9(0.3)   221.3(1.3)

$\Delta_{hyd}H$ was measured in *n*-hexane solution against 1-hexene as a standard. The value given was determined from the $\Delta_{hyd}H$ of a mixture of the isomers and glc determination of the ratio of isomers in the sample.

Fang, W.; Rogers, D. W. *J. Org. Chem.* **1992**, *57*, 2294–2297.

(5)   1,4-Hexadiene, (Z)
        cis-1,4-Hexadiene
        298 K   58.4(0.4)   244.3(1.7)

$\Delta_{hyd}H$ was measured in *n*-hexane solution against 1-hexene as a standard. The value given is the arithmetic mean of two experiments. Uncertainties are the arithmetic mean of the uncertainties for each experiment which, in turn, are twice the standard deviation from the mean for four degrees of freedom.

Fang, W.; Rogers, D. W. *J. Org. Chem.* **1992**, *57*, 2294–2297.

(5')   1,4-Hexadiene, (E)
        298 K   57.6(0.4)   241.0(1.7)

$\Delta_{hyd}H$ was measured in *n*-hexane solution against 1-hexene as a standard. The value given is the arithmetic mean of two experiments. Uncertainties are the arithmetic mean of the uncertainties for each experiment which, in turn, are twice the standard deviation from the mean for four degrees of freedom.

Fang, W.; Rogers, D. W. *J. Org. Chem.* **1992**, *57*, 2294–2297.

(6) 1,5-Hexadiene
     355 K   60.5(0.2)   253.3(0.8)

$\Delta_{hyd}H$ was measured in the gas phase. Compare the following two entries.

Kistiakowsky, G. B.; Ruhoff, J. R.; Smith, H. A.; Vaughan, W. E. *J. Amer. Chem. Soc.* **1936**, *58*, 146–153.

(6') 1,5-Hexadiene
      298 K   60.2(0.4)   251.7(1.5)

$\Delta_{hyd}H$ was measured in glacial acetic acid.

Turner, R. B.; Mallon, B. J.; Tichy, M.; Doering, W. von E.; Roth. W. R.; Schroeder, G. *J. Amer. Chem. Soc.* **1973**, *95*, 8605–8609.

(6") 1,5-Hexadiene
      298 K   60.3(0.4)   252.3(1.7)

$\Delta_{hyd}H$ was measured in *n*-hexane solution against 1-hexene as a standard. The value given is the arithmetic mean of two experiments. Uncertainties are the arithmetic mean of the uncertainties for each experiment which, in turn, are twice the standard deviation from the mean for four degrees of freedom.

Fang, W.; Rogers, D. W. *J. Org. Chem.* **1992**, *57*, 2294–2297.

(7) 2,4-Hexadiene, (Z,Z)
     cis,cis-2,4-Hexadiene
     298 K   52.4(0.4)   219.2(1.7)

$\Delta_{hyd}H$ was measured in *n*-hexane solution against 1-hexene as a standard. The value given is the arithmetic mean of two experiments. Uncertainties are the arithmetic mean of the uncertainties for each experiment which, in turn, are twice the standard deviation from the mean for four degrees of freedom.

Fang, W.; Rogers, D. W. *J. Org. Chem.* **1992**, *57*, 2294–2297.

(8) 2,4-Hexadiene, (Z,E)
     cis,trans-2,4-Hexadiene
     298 K   51.4(0.4)   215.0(1.7)

$\Delta_{hyd}H$ was measured in *n*-hexane solution. Read "2,4" for "1,4" in the original paper.

Fang, W.; Rogers, D. W. *J. Org. Chem.* **1992**, *57*, 2294–2297.

(9)   Hexa-2,4-diene, (E,E)
       trans,trans-2,4-Hexadiene
       298 K   50.5(0.4)   211.3(1.7)

$\Delta_{hyd}H$ was measured in n-hexane solution. The Z-E stabilization-destabilization energy relationship is quite evident in the sequence 52.4, 51.4, 50.5 seen in the last three entries.

Fang, W.; Rogers, D. W. *J. Org. Chem.* **1992**, *57*, 2294–2297.

(10)   2,3-Dimethylbuta-1,3-diene
       355 K   53.87(0.2)   253.3(0.6)

Compare with 1,3-butadiene. The stabilizing effect of two α methyl groups is nearly as great as 1,3-conjugation and the two stabilizing effects are additive.

Dolliver, M. A.; Gresham, T. L.; Kistiakowsky, G. B.; Vaughan, W. E. *J. Amer. Chem. Soc.* **1937**, *59*, 831–841.

(11)   Cyclohexene
       355 K   28.59(0.1)   119.6(0.4)

Compare the next 5 entries. Six sets of results obtained by 5 different research groups over 47 years have an average $\Delta_{hyd}H$ = -28.5 ± 0.13 kcal mol^{-1}. Two data sets have been corrected from the liquid phase to the gas phase (Fuchs, R.; Peacock, L. A., *J. Phys. Chem.* **1979**, *83*, 1975–1978).

Kistiakowsky, G. B.; Ruhoff, J. R.; Smith, H. A.; Vaughan, W. E. *J. Amer. Chem. Soc.* **1936**, *58*, 137–145.

(11') Cyclohexene
       298 K   27.1(0.1)   113.4(0.4)

$\Delta_{hyd}H$ was measured in acetic acid solution and corrected to 28.6 kcal mol^{-1} = 119 kJ mol^{-1} in the gas phase (Fuchs, R.; Peacock, L. A., *J. Phys. Chem.* **1979**, *83*, 1975–1978).

Turner, R. B.; Meador, W. R.; Winkler, R. E. *J. Amer. Chem. Soc.* **1957**, *79*, 4116–4121.

(11") Cyclohexene
       298 K   26.9(0.6)   113.0(2.5)

Hydrogenation was carried out in glacial acetic acid. Correction to the gas phase by the method of Fuchs and Peacock (above) yields 28.4 kcal mol^{-1} = 119.0 kJ mol^{-1}.

Rogers, D. W.; Mc Lafferty, F. J., *Tetrahedron* **1971**, *27*, 3765–3775.

(11''') Cyclohexene
       298 K   28.4(0.1)   118.8(0.4)

Roth, W. R.; Lennartz, H.-W. *Chem. Ber.* **1980**, *113*, 1806–1817.

(11'''') Cyclohexene
298 K   28.35(0.05)   118.6(0.2)

The determination was carried out in a thermoneutral solvent using a microprocessor-controlled electrical calibration circuit programmed to deliver a 4.136 J energy pulse to the calorimeter over a heating time of 5.273 s (Rogers, D. W.; Munoz-Hresko, B., *Mikrochimica Acta [Wien]* **1984** *I* 417–425). Uncertainties are 95% confidence limits for 9 replicate determinations. Results are reported in four significant figures because the point of the work is precision testing.

Rogers, D. W.; Munoz-Hresko, B., *Mikrochimica Acta [Wien]* **1984** *II, 427*–435.

(11''''') Cyclohexene
298 K   28.30(0.07)   118.4(0.3)

The determination was carried out in a thermoneutral solvent using a microprocessor-controlled electrical calibration circuit programmed to deliver a 4.136 J energy pulse to the calorimeter over a heating time of 5.273 s (Rogers, D. W.; Munoz-Hresko, B., *Mikrochimica Acta [Wien]* **1984** *I* 417–425). Uncertainties are 95% confidence limits for 9 replicate determinations. Results are reported in four significant figures because the point of the work is precision testing.

Rogers, D. W.; Munoz-Hresko, B., *Mikrochimica Acta [Wien]* **1984** *II, 427*–435.

(12) Methylenecyclopentane
298 K   26.8(0.1)   112.2(0.4)

$\Delta_{\mathrm{hyd}} H$ was measured in acetic acid solution. This result is part of a preliminary study on endocyclic-exocyclic stability that was completed and augmented in Turner, R. B.; Garner, R. H. *J. Amer. Chem. Soc.* **1958**, *80*, 1424–1430.

Turner, R. B.; Garner, R. H. *J. Amer. Chem. Soc.* **1957**, *79*, 253 (one page).

(12') Methylenecyclopentane
298 K   26.9(0.1)   112.5(0.4)

$\Delta_{\mathrm{hyd}} H$ was measured in acetic acid solution. This result is part of a more general study of exocyclic-endocyclic stability started in Turner, R. B.; Garner, R. H. *J. Amer. Chem. Soc.* **1957**, *79*, 253 (one page). Steric and electronic stabilization is discussed in detail.

Turner, R. B.; Garner, R. H. *J. Amer. Chem. Soc.* **1958**, *80*, 1424–1430.

(12'') Methylenecyclopentane
298 K   27.7(0.1)   115.9(0.4)

$\Delta_{hyd}H$ was measured in *n*-hexane solution. See also, Turner, R. B.; Garner, R. H. *J. Amer. Chem. Soc.* **1958**, *80*, 1424–1430.

Allinger, N. L.; Dodziuk, H.; Rogers, D. W.; Naik, S. N., *Tetrahedron* **1982**, *38*, 1593–1597.

(13)   1-Methylcyclopentene
        298 K   23.0(0.1)   96.3(0.4)

$\Delta_{hyd}H$ was measured in acetic acid solution. Steric and electronic stabilization is discussed in detail.

Turner, R. B.; Garner, R. H. *J. Amer. Chem. Soc.* **1957**, 79, 253 (one page).

(13')   1-Methylcyclopentene
        298 K   24.2(0.1)   101.3(0.4)

$\Delta_{hyd}H$ was measured in *n*-hexane solution. See also, Turner, R. B.; Garner, R. H. *J. Amer. Chem. Soc.* **1958**, *80*, 1424–1430.

Allinger, N. L.; Dodziuk, H.; Rogers, D. W.; Naik, S. N., *Tetrahedron* **1982**, *38*, 1593–1597.

(14)   3-Methylcyclopentene
        298 K   27.6(0.1)   115.4(0.4)

$\Delta_{hyd}H$ was measured in *n*-hexane solution.

Allinger, N. L.; Dodziuk, H.; Rogers, D. W.; Naik, S. N., *Tetrahedron* **1982**, *38*, 1593–1597.

(15)   1,2-Dimethylcyclobutene
        298 K   26.4(0.2)   110.3(0.8)

$\Delta_{hyd}H$ was measured in acetic acid.

Turner, R. B.; Goebel, P.; Mallon, B. J.; von E. Doering, W.; Coburn Jr., J. F.; Pomerantz, M. *J. Amer. Chem. Soc.* **1968**, *90*, 4315–4322.

(16)   1,3-Dimethylbicyclo[1.1.0]butane
        298 K   40.6(0.4)   169.8(1.7)

$\Delta_{hyd}H$ was measured in diethyl carbitol. More than one equivalent of H_2 was taken up. The value given is per mol of H_2.

Turner, R. B.; Goebel, P.; Mallon, B. J.; von E. Doering, W.; Coburn Jr., J. F.; Pomerantz, M. *J. Amer. Chem. Soc.* **1968**, *90*, 4315–4322. See also Turner, R. B.; Goebel, P.; von E. Doering, W.; Coburn Jr., J. F. *Tetrahedron Letters* **1965**, *15*, 997–1002.

(17) Ethylbuta-1,3-diene
298 K   56.3   235.6

One measurement was made, hence no experimental uncertainty is given.

Roth, W. R.; Adamczak, O.; Breuckman, R.; Lennartz, H.-W.; Boese, R. *Chem. Ber.* **1991**, *124*, 2499–2521.

(18) Bicyclo[2.2.0]hexane
298 K   59.3(0.2)   248.1(0.8)

The product was cyclohexane.

Roth, W. R.; Klaerner F.-G.; Lennartz, H.-W. *Chem. Ber.* **1980**, *113*, 1818–1829.

(19) 1-Cyclopropylpropene
298 K   63.0(0.2)   263.6(0.8)

The reaction was carried out in isooctane. The reaction product was *n*-hexane.

Roth, W. R.; Kirmse, W.; Hoffmann, W.; Lennartz, H.-W. *Chem. Ber.* **1982**, *115*, 2508–2515.

(20) 1-Vinyl-2-methylcyclopropane
298 K   64.5(0.1)   269.9(0.4)

The reaction was carried out in isooctane. The reaction product was *n*-hexane.

Roth, W. R.; Kirmse, W.; Hoffmann, W.; Lennartz, H.-W. *Chem. Ber.* **1982**, *115*, 2508–2515.

C_6H_{12}

(1) 1-Hexene
298 K   30.1(0.4)   125.9(1.7)

$\Delta_{hyd}H$ was measured in glacial acetic acid and corrected by a separate heat of solution measurement experiment for *n*-hexane. The value given is for a liquid → liquid reaction.

Flitcroft, T. L,; Skinner, H. A.; Whiting, M. C., *Trans. Faraday Soc.* **1957**, *53*, 784–790.

(1')   1-Hexene
       298 K   30.2(0.2)   126.6(0.8)

$\Delta_{hyd}H$ was measured in glacial acetic acid and corrected by a separate heat of solution measurement experiment for n-hexane. The value given for the liquid → liquid reaction.

Skinner, H. A.; Snelson A., *Trans. Faraday Soc.* **1959**, *55*, 404–407.

(1")   1-Hexene
       299.1   29.90(0.3)   125.1(1.3)

Hydrogenation was carried out in an oscillatory calorimeter using glacial acetic acid as the calorimeter fluid. Results are corrected for $\Delta_{hyd}H$ of the product in acetic acid.

Bretschneider, E.; Rogers, D. W. *Mikrochemica. Acta [Wien]* **1970**, 482–490.

(1''') 1-Hexene
       298 K   30.2(0.2)   126.2(0.8)

Hydrogenation was carried out in n-hexane. Value differs slightly from the original paper due to a revised standard $\Delta_{hyd}H$, Rogers, D. W., *J. Phys. Chem.* **1979**, *83*, 2430 (one page).

Rogers, D. W.; Papadimetriou. P. M.; Siddiqui, N. A., *Mikrochimica Acta [Wien]* **1975** II, 389–400.

(1'''') 1-Hexene
       298 K   30.3(0.3)   126.8(1.3)

Hydrogenation was carried out in n-hexane. Value differs slightly from the original paper due to a revised standard $\Delta_{hyd}H$, Rogers, D. W., *J. Phys. Chem.* **1979**, *83*, 2430 (one page).

Rogers, D. W.; Skanupong, S., *J. Phys. Chem.* **1974**, *78*, 2569–2572.

(1''''') 1-Hexene
       298 K   30.2(0.2)   126.6(0.8)

Rogers, D. W., *J. Phys. Chem.* **1979**, *83*, 2430 (one page).

(1'''''') 1-Hexene
       298 K   30.3(0.1)   126.7(0.5)

The value given is the arithmetic mean of two experiments. Uncertainties are the arithmetic mean of the uncertainties for each experiment which, in turn, are twice the standard deviation from the mean for four degrees of freedom.

Rogers, D. W., Crooks, E. Dejroongruang, K. *J. Chem. Thermodynam.* **1987**, *19*, 1209–1215.

(2) 2-Hexene, (Z)
     cis-2-Hexene
     298 K   27.7(0.2)   115.8(0.8)

Hydrogenation was carried out in n-hexane. Value differs slightly from the original paper due to a revised standard $\Delta_{hyd}H$, Rogers, D. W., *J. Phys. Chem.* **1979**, *83*, 2430 (one page).

Rogers, D. W.; Papadimetriou. P. M.; Siddiqui, N. A., *Mikrochimica Acta [Wien]* **1975** II, 389–400.

(2') 2-Hexene, (Z)
      cis-2-Hexene
      298 K   28.44(0.19)   119.0(0.78)

This work was undertaken to investigate the precision of the method, hence the results are presented with an extra significant digit. The entry above represents the 95% confidence limits of four separate experiments consisting of 9 runs each, 36 hydrogenations in all. The results do not support the notion that hydrogen thermochemistry is inherently inaccurate because of kinetic lag.

Rogers, D. W.; Crooks, E. L., *J. Chem. Thermodynam.* **1983**, *15*, 1087–1092.

(2") 2-Hexene, (Z)
      cis-2-Hexene
      298 K   28.6(0.2)   119.5(0.7)

The value given is the arithmetic mean of two experiments. Uncertainties are the arithmetic mean of the uncertainties for each experiment which, in turn, are twice the standard deviation from the mean for four degrees of freedom.

Rogers, D. W., Crooks, E. Dejroongruang, K. *J. Chem. Thermodynam.* **1987**, *19*, 1209–1215.

(3) 2-Hexene, (E)
     298 K   27.3(0.3)   113.8(1.3)

Hydrogenation was carried out in *n*-hexane. Value differs slightly from the original paper due to a revised standard $\Delta_{hyd}H$, Rogers, D. W., *J. Phys. Chem.* **1979**, *83*, 2430 (one page).

Rogers, D. W.; Papadimetriou. P. M.; Siddiqui, N. A., *Mikrochimica Acta [Wien]* **1975** II, 389–400.

(3')   2-Hexene, (E)
        trans-2-Hexene
        298 K   27.48(0.20)   115.0(0.83)

This work was undertaken to investigate the precision of the method, hence the results are presented with an extra significant digit. The entry above represents the 95% confidence limits of four separate experiments consisting of 9 runs each, 36 hydrogenations in all. The results do not support the notion that hydrogen thermochemistry is inherently inaccurate because of kinetic lag.

Rogers, D. W.; Crooks, E. L., *J. Chem. Thermodynam.* **1983**, *15*, 1087–1092.

(3")   2-Hexene, (E)
        trans-2-Hexene
        298 K   27.7(0.1)   116.1(0.5)

The value given is the arithmetic mean of two experiments. Uncertainties are the arithmetic mean of the uncertainties for each experiment which, in turn, are twice the standard deviation from the mean for four degrees of freedom.

Rogers, D. W., Crooks, E. Dejroongruang, K. *J. Chem. Thermodynam.* **1987**, *19*, 1209–1215.

(4)   3-Hexene, (Z)
        cis-3-Hexene
        298 K   29.2(0.3)   122.1(1.3)

Hydrogenation was carried out in *n*-hexane. Value differs slightly from the original paper due to a revised standard $\Delta_{hyd}H$, Rogers, D. W., *J. Phys. Chem.* **1979**, *83*, 2430 (one page).

Rogers, D. W.; Papadimetriou. P. M.; Siddiqui, N. A., *Mikrochimica Acta [Wien]* **1975** II, 389–400.

(4')   3-Hexene, (Z)
        cis-3-Hexene
        298 K   29.3(0.3)   122.6(1.3)

Hydrogenation was carried out in *n*-hexane. Value differs slightly from the original paper due to a revised standard $\Delta_{hyd}H$, Rogers, D. W., *J. Phys. Chem.* **1979**, *83*, 2430 (one page).

Rogers, D. W.; Siddiqui, N. A., *J. Phys. Chem.* **1975**, *79*, 574–577.

(4")   3-Hexene, (Z)
        cis-3-Hexene
        298 K   28.94(0.20)   121.1(0.85)

This work was undertaken to investigate the precision of the method, hence the results are presented with an extra significant digit. The entry above represents the 95% confidence limits of four separate experiments consisting of 9 runs each, 36 hydrogenations in all. The results do not support the notion that hydrogen thermochemistry is inherently inaccurate because of kinetic lag.

Rogers, D. W.; Crooks, E. L., *J. Chem. Thermodynam.* **1983**, *15*, 1087–1092.

(4''')   3-Hexene, (Z)
         cis-3-hexene
         298 K   29.02(0.07)   121.4(0.3)

The determination was carried out in a thermoneutral solvent using a microprocessor-controlled electrical calibration circuit programmed to deliver a 4.136 J energy pulse to the calorimeter over a heating time of 5.273 s (Rogers, D. W.; Munoz-Hresko, B., *Mikrochimica Acta [Wien]* **1984** *I* 417–425. Uncertainties are 95% confidence limits for 9 replicate determinations. Results are reported in four significant figures because the point of the work is precision testing.

Rogers, D. W.; Munoz-Hresko, B., *Mikrochimica Acta [Wien]* **1984** *II, 427*–435.

(4'''')   3-Hexene, (Z)
          cis-3-Hexene
          298 K   28.92(0.12)   121.0(0.5)

The determination was carried out in a thermoneutral solvent using a microprocessor-controlled electrical calibration circuit programmed to deliver a 4.136 J energy pulse to the calorimeter over a heating time of 5.273 s (Rogers, D. W.; Munoz-Hresko, B., *Mikrochimica Acta [Wien]* **1984** *I* 417–425. Uncertainties are 95% confidence limits for 9 replicate determinations. Results are reported in four significant figures because the point of the work is precision testing.

Rogers, D. W.; Munoz-Hresko, B., *Mikrochimica Acta [Wien]* **1984** *II, 427*–435.

(4''''') 3-Hexene(Z)
          cis-3-Hexene
          298 K   29.1(0.1)   121.6(0.3)

The value given is the arithmetic mean of two experiments. Uncertainties are the arithmetic mean of the uncertainties for each experiment which, in turn, are twice the standard deviation from the mean for four degrees of freedom.

Rogers, D. W., Crooks, E. Dejroongruang, K. *J. Chem. Thermodynam.* **1987**, *19*, 1209–1215.

(5)   3-Hexene, (E)
        trans-3-hexene
        298 K   26.9(0.4)   112.3(1.7)

Hydrogenation was carried out in *n*-hexane. Value differs slightly from the original paper due to a revised standard $\Delta_{hyd}H$, Rogers, D. W., *J. Phys. Chem.* **1979**, *83*, 2430 (one page).

Rogers, D. W.; Papadimetriou. P. M.; Siddiqui, N. A. *Mikrochimica Acta [Wien]* **1975** II, 389–400.

(5')   3-Hexene, (E)
        trans-3-Hexene
        298 K   28.18(0.2)   117.9(0.8)

This work was undertaken to investigate the precision of the method, hence the results are presented with an extra significant digit. The entry above represents the 95% confidence limits of four separate experiments consisting of 9 runs each, 36 hydrogenations in all. The results do not support the notion that hydrogen thermochemistry is inherently inaccurate because of kinetic lag.

Rogers, D. W.; Crooks, E. L. *J. Chem. Thermodynam.* **1983**, *15*, 1087–1092.

(5")   3-Hexene, (E)
        trans-2-Hexene
        298 K   27.7(0.2)   117.9(0.7)

The value given is the arithmetic mean of two experiments. Uncertainties are the arithmetic mean of the uncertainties for each experiment which, in turn, are twice the standard deviation from the mean for four degrees of freedom

Rogers, D. W., Crooks, E. Dejroongruang, K. *J. Chem. Thermodynam.* **1987**, *19*, 1209–1215.

(6)   2-Methylpent-1-ene
        298 K   27.8(0.1)   116.3(0.4)

The value given is the arithmetic mean of two experiments. Uncertainties are the arithmetic mean of the uncertainties for each experiment which, in turn, are twice the standard deviation from the mean for four degrees of freedom.

Rogers, D. W., Crooks, E. Dejroongruang, K. *J. Chem. Thermodynam.* **1987**, *19*, 1209–1215.

(7) 3-Methylpent-1-ene
     298 K   29.8(0.1)   124.6(0.5)

The value given is the arithmetic mean of two experiments. Uncertainties are the arithmetic mean of the uncertainties for each experiment which, in turn, are twice the standard deviation from the mean for four degrees of freedom.

Rogers, D. W., Crooks, E. Dejroongruang, K. *J. Chem. Thermodynam.* **1987**, *19*, 1209–1215.

(8) 4-Methylpent-1-ene
     298 K   30.3(0.1)   126.7(0.4)

The value given is the arithmetic mean of two experiments. Uncertainties are the arithmetic mean of the uncertainties for each experiment which, in turn, are twice the standard deviation from the mean for four degrees of freedom.

Rogers, D. W., Crooks, E. Dejroongruang, K. *J. Chem. Thermodynam.* **1987**, *19*, 1209–1215.

(9) 2-Methylpent-2-ene
     298 K   26.7(0.2)   111.6(0.7)

The value given is the arithmetic mean of two experiments. Uncertainties are the arithmetic mean of the uncertainties for each experiment which, in turn, are twice the standard deviation from the mean for four degrees of freedom.

Rogers, D. W., Crooks, E. Dejroongruang, K. *J. Chem. Thermodynam.* **1987**, *19*, 1209–1215.

(10) 3-Methylpent-2-ene, (Z)
     cis-3-Methylpent-2-ene
     298 K   26.4(0.1)   110.6(0.4)

The value given is the arithmetic mean of two experiments. Uncertainties are the arithmetic mean of the uncertainties for each experiment which, in turn, are twice the standard deviation from the mean for four degrees of freedom.

Rogers, D. W., Crooks, E. Dejroongruang, K. *J. Chem. Thermodynam.* **1987**, *19*, 1209–1215.

(11) 3-Methylpent-2-ene, (E)
     trans-3-Methylpent-2-ene
     298 K   26.3(0.1)   110.1(0.6)

The value given is the arithmetic mean of two experiments. Uncertainties are the arithmetic mean of the uncertainties for each experiment which, in turn, are twice the standard deviation from the mean for four degrees of freedom.

Rogers, D. W., Crooks, E. Dejroongruang, K. *J. Chem. Thermodynam.* **1987**, *19*, 1209–1215.

(12) 4-Methylpent-2-ene, (Z)
 4-Methyl-*cis*-2-pentene
 cis-3-Methylpent-2-ene
 298 K   27.3(0.1)   114.3(0.4)

$\Delta_{hyd}H$ was measured in acetic acid

Turner, R. B.; Nettleton, D. E.; Perelman, M. *J. Amer. Chem. Soc.* **1958**, *80*, 1430–1433.

(12') 4-Methylpent-2-ene, (Z)
 4-Methyl-*cis*-2-pentene
 cis-3-Methylpent-2-ene
 298 K   27.9(0.1)   116.9(0.4)

The value given is the arithmetic mean of two experiments. Uncertainties are the arithmetic mean of the uncertainties for each experiment which, in turn, are twice the standard deviation from the mean for four degrees of freedom.

Rogers, D. W., Crooks, E. Dejroongruang, K. *J. Chem. Thermodynam.* **1987**, *19*, 1209–1215.

(13) 4-Methylpent-2-ene, (E)
 4-Methyl-*trans*-2-pentene
 trans-4-Methyl-2-pentene
 298 K   26.4(0.1)   110.4(0.4)

$\Delta_{hyd}H$ was measured in acetic acid.

Turner, R. B.; Nettleton, D. E.; Perelman, M. *J. Amer. Chem. Soc.* **1958**, *80*, 1430–1433.

(13') 4-Methylpent-2-ene, (E)
 4-Methyl-*trans*-2-pentene
 trans-3-Methylpent-2-ene
 298 K   27.3(0.1)   114.2(0.6)

The value given is the arithmetic mean of two experiments. Uncertainties are the arithmetic mean of the uncertainties for each experiment which, in turn, are twice the standard deviation from the mean for four degrees of freedom.

Rogers, D. W., Crooks, E. Dejroongruang, K. *J. Chem. Thermodynam.* **1987**, *19*, 1209–1215.

(14) Ethylbut-1-ene
      3-Methylene-1-pentane
      298 K   27.7(0.1)   115.8(0.4)

The value given is the arithmetic mean of two experiments. Uncertainties are the arithmetic mean of the uncertainties for each experiment which, in turn, are twice the standard deviation from the mean for four degrees of freedom.

Rogers, D. W., Crooks, E. Dejroongruang, K. *J. Chem. Thermodynam.* **1987**, *19*, 1209–1215.

(15) 2,3-Dimethylbut-1-ene
      unsym-Methylisopropylethylene
      298 K   28.00(0.1)   117.2(0.4)

Kistiakowsky, G. B.; Ruhoff, J. R.; Smith, H. A.; Vaughan, W. E. *J. Amer. Chem. Soc.* **1936**, *58*, 137–145.

(15') 2,3-Dimethylbut-1-ene
       298 K   27.8(0.1)   116.3(0.6)

The value given is the arithmetic mean of two experiments. Uncertainties are the arithmetic mean of the uncertainties for each experiment which, in turn, are twice the standard deviation from the mean for four degrees of freedom.

Rogers, D. W., Crooks, E. Dejroongruang, K. *J. Chem. Thermodynam.* **1987**, *19*, 1209–1215.

(16) 3,3-Dimethylbut-1-ene
      t-Butylethylene
      355 K   30.3(0.2)   126.9(0.8)

Dolliver, M. A.; Gresham, T. L.; Kistiakowsky, G. B.; Vaughan, W. E. *J. Amer. Chem. Soc.* **1937**, *59*, 831–841.

(16') 3,3-Dimethylbut-1-ene
       298 K   30.1(0.1)   125.8(0.5)

The value given is the arithmetic mean of two experiments. Uncertainties are the arithmetic mean of the uncertainties for each experiment which, in turn, are twice the standard deviation from the mean for four degrees of freedom.

Rogers, D. W., Crooks, E. Dejroongruang, K. *J. Chem. Thermodynam.* **1987**, *19*, 1209–1215.

(17) 2,3-Dimethylbut-2-ene
      Tetramethylethylene
      298 K   26.63(0.1)   111.4(0.4)

Kistiakowsky, G. B.; Ruhoff, J. R.; Smith, H. A.; Vaughan, W. E. *J. Amer. Chem. Soc.* **1936**, *58*, 137–145.

(17') 2,3-Dimethylbut-2-ene
       298 K   26.0(0.1)   108.7(0.5)

The value given is the arithmetic mean of two experiments. Uncertainties are the arithmetic mean of the uncertainties for each experiment which, in turn, are twice the standard deviation from the mean for four degrees of freedom.

Rogers, D. W., Crooks, E. Dejroongruang, K. *J. Chem. Thermodynam.* **1987**, *19*, 1209–1215.

C_7H_8

(1) 1,3,5-Cycloheptatriene
     Tropylidene
     355 K   72.8(0.3) 304.8(1.3)

Conn, J. B.; Kistiakowsky, G. B.; Smith, E. A. *J. Amer. Chem. Soc.* **1939**, *61*, 1868–1876.

(1') 1,3,5-Cycloheptatriene
      Cyclohepta-1,3,5-triene
      298 K   70.5(0.4)   294.9(1.7)

$\Delta_{hyd}H$ was measured in glacial acetic acid solution. The discrepancy between the value measured in acetic acid and the gas-phase value is more pronounced for the triene and, in this case, it is nearly linear in the number of double bonds, 3(0.7 kcal mol^{-1}). In general, the difference is greater for polyunsaturates but it is not linear in the number of double bonds. See also Turner, R. B.; Mallon, B. J.; Tichy, M.; Doering, W. von E.; Roth. W. R.; Schroeder, G. *J. Amer. Chem. Soc.* **1973**, *95*, 8605–8609.

Turner, R. B.; Meador, W. R.; Winkler, R. E. *J. Amer. Chem. Soc.* **1957**, *79*, 4116–4121.

(1") 1,3,5-Cycloheptatriene
      298 K   72.8(0.1)   304.6(0.4)

The result is corrected for the difference in enthalpies of solution of the reactant and the product.

Roth, W. R.; Klaerner, F.-G.; Grimme, W.; Koeser, H. G.; Busch, R.; Muskulus, B.; Breuckmann, R.; Scholz, B. P.; Lennartz, H.-W. *Chem. Ber.* **1983**, *116*, 2717–2737.

(2)   Bicyclo[2.2.1]hepta-2,5-diene
        Norbornadiene
        298 K   68.1(0.4)   285.0(1.7)

$\Delta_{hyd}H$ was measured in acetic acid solution.

Turner, R. B.; Meador, W. R.; Winkler, R. E. *J. Amer. Chem. Soc.*. **1957**, *79*, 4116–4121.

(2')   Bicyclo[2.2.1]hepta-2,5-diene
        Norbornadiene
        298 K   69.8(0.4)   291.9(1.7)

Rogers, D. W.; Choi, L. S.; Girellini, R. S.; Holmes, T. J.; Allinger, N. L., *J. Phys. Chem.* **1980**, *84*, 1810–1814.

(2")   Bicyclo[2.2.1]hepta-2,5-diene
        Norbornadiene
        298 K   70.8(0.3)   296.2(1.3)

$\Delta_{hyd}H$ was measured in isooctane solution and corrected for $\Delta_{hyd}H$ and $\Delta_{sol'n}H$ according to the difference in Kovats indices of the reactant and product.

Doering, W. v. E; Roth, W. R.; Breuckman, R.; Figge, L.; Lennartz, H.-W.; Fessner, W.-D.; Prinzbach. H. *Chem. Ber.* **1988**, *121*, 1–9.

(3)   Tetracyclo[2.2.1.0.^{2,6}0^{3,5}]heptane
        Quadricyclane
        Quadricyclene
        298 K   92.0(0.5)   385.1(2.1)

$\Delta_{hyd}H$ was measured in acetic acid.

Turner, R. B.; Goebel, P.; Mallon, B. J.; von E. Doering, W.; Coburn Jr., J. F.; Pomerantz, M. *J. Amer. Chem. Soc.* **1968**, *90*, 4315–4322.

(3')   Tetracyclo[2.2.1.0.^{2,6}0^{3,5}]heptane
        Quadricyclane
        Quadricyclene
        298 K   66.4(0.4)   277.9(2.0)

Hydrogenation$_{2,6}$ carried out over Pd catalyst, was incomplete, yielding tricyclo[2.2.1.0^{2,6}]heptane (nortricyclane) as the major product.

Rogers, D. W.; Choi, L. S.; Girellini, R. S.; Holmes, T. J.; Allinger, N. L., *J. Phys. Chem.* **1980**, *84*, 1810–1814.

(3") Tetracyclo[2.2.1.0.^{2,6}0^{3,5}]heptane
Quadricyclane
Quadricyclene
298 K   91.9(0.6)   385(2.5)

Hydrogenation, over Pt catalyst, was complete, yielding norbornane.

Rogers, D. W.; Choi, L. S.; Girellini, R. S.; Holmes, T. J.; Allinger, N. L., *J. Phys. Chem.* **1980**, *84*, 1810–1814.

(3''') Tetracyclo[2.2.1.0.^{2,6}0^{3,5}]heptane
Quadricyclane
Quadricyclene
298 K   91.9(0.4)   385(1.7)

Hydrogenation was over Rh catalyst, yielding norbornane.

Rogers, D. W.; Choi, L. S.; Girellini, R. S.; Holmes, T. J.; Allinger, N. L., *J. Phys. Chem.* **1980**, *84*, 1810–1814.

(4) Spiro[2.4]cyclohepta-1,3-diene
Cyclopropylcyclopenta-1,3-diene
298 K   87.1(0.2)   364.4(0.8)

Roth, W. R.; Adamczak, O.; Breuckman, R.; Lennartz, H.-W.; Boese, R. *Chem. Ber.* **1991**, *124*, 2499–2521.

(5) Ethylidinecyclopenta-1,3-diene
298 K   74.5(0.1)   311.7(0.4)

Roth, W. R.; Adamczak, O.; Breuckman, R.; Lennartz, H.-W.; Boese, R. *Chem. Ber.* **1991**, *124*, 2499–2521.

(6) *exo*-Tricyclo[3.2.0.0^{2,4}]hept-6-ene
298 K   91.1(0.1)   381.2(0.4)

The product is bicyclo[3.2.0]heptane. The result is corrected for a small amount of methylcyclohexane in the product.

Roth, W. R.; Klaerner, F.-G.; Grimme, W.; Koeser, H. G.; Busch, R.; Muskulus, B.; Breuckmann, R.; Scholz, B. P.; Lennartz, H.-W. *Chem. Ber.* **1983**, *116*, 2717–2737.

C_7H_{10}

(1)   1,2,6-Heptatriene

298 K   98.0(0.2)   410.0(0.8)

The results were corrected for enthalpies of solution, but not for heats of vaporization.

Roth, W. R.; Wollweber, D.; Offerhaus, R.; Rekowski, V.; Lennartz, H.-W. Sustmann, R.; Muller, W. *Chem. Ber.* **1993**, *126*, 2701–2715.

(2)   3-Methylenehexa-1,5-diene
298 K   84.8(0.2)   354.8(0.8)

Roth, W. R.; Wollweber, D.; Offerhaus, R.; Rekowski, V.; Lennartz, H.-W. Sustmann, R.; Muller, W. *Chem. Ber.* **1993**, *126*, 2701–2715.

(3)   3-Vinylpenta-1,3-diene
298 K   83.3(0.1)   348.5(0.4)

Roth, W. R.; Adamczak, O.; Breuckman, R.; Lennartz, H.-W.; Boese, R. *Chem. Ber.* **1991**, *124*, 2499–2521.

(4)   Cycloheptadiene
355 K   51.3(0.2)   214.5(0.8)

The isomer is not identified in the paper, but it was evidently the 1,3-diene.

Conn, J. B.; Kistiakowsky, G. B.; Smith, E. A. *J. Am. Chem. Soc.* **1939**, *61*, 1868–1876.

(4')   1,3-Cycloheptadiene
298 K   49.9(0.1)   208.9(0.4)

Turner, R. B.; Mallon, B. J.; Tichy, M.; Doering, W. von E.; Roth. W. R.; Schroeder, G. *J. Amer. Chem. Soc.* **1973**, *95*, 8605–8609.

(5)   1,4-Cycloheptadiene
298 K   55.9(0.1)   233.8(0.4)

The hydrogenation was carried out in acetic acid solution.

Turner, R. B.; Mallon, B. J.; Tichy, M.; Doering, W. von E.; Roth. W. R.; Schroeder, G. *J. Amer. Chem. Soc.* **1973**, *95*, 8605–8609.

(6)   Bicyclo[2.2.1]hept-2-ene
        Norbornene
298 K   33.1(0.2)   138.6(0.8)

Experimental Results 57

$\Delta_{hyd}H$ was measured in acetic acid solution.

Turner, R. B.; Meador, W. R.; Winkler, R. E. *J. Amer. Chem. Soc.* **1957**, *79*, 4116–4121.

(6')   Bicyclo[2.2.1]hept-2-ene
        Norbornene
        298 K   33.8(0.3)   142(1.3)

Rogers, D. W.; Choi, L. S.; Girellini, R. S.; Holmes, T. J.; Allinger, N. L., *J. Phys. Chem.* **1980**, *84*, 1810–1814.

(6")   Bicyclo[2.2.1]hept-2-ene
        Norbornene
        298 K   32.8(0.1)   137.2(0.4)

$\Delta_{hyd}H$ was measured in isooctane solution and corrected for $\Delta_{vap}H$ and for $\Delta_{sol'n}H$ according to the difference in Kovats indices of the reactant and product.

Doering, W. v. E; Roth, W. R.; Breuckman, R.; Figge, L.; Lennartz, H.-W.; Fessner, W.-D.; Prinzbach. H. *Chem. Ber.* **1988**, *121*, 1–9.

(7)   Bicyclo[4.1.0]hept-2-ene
        298 K   64.2(0.1)   268.6(0.4)

Hydrogenation yields a mixture of products, mainly cycloheptane and toluene.

Roth, W. R.; Klaerner, F.-G.; Siepert, G.; Lennartz, H.-W. *Chem. Ber.* **1992**, *125*, 217–224.

(8)   Bicyclo[4.1.0]hept-3-ene
        298 K   63.3(0.1)   264.8(0.4)

Hydrogenation yields a mixture of products, mainly cycloheptane and toluene.

Roth, W. R.; Klaerner, F.-G.; Siepert, G.; Lennartz, H.-W. *Chem. Ber.* **1992**, *125*, 217–224.

(9)   1,1-Divinylcyclopropane
        298 K   93.4(0.2)   390.8(0.8)

Roth, W. R.; Adamczak, O.; Breuckman, R.; Lennartz, H.-W.; Boese, R. *Chem. Ber.* **1991**, *124*, 2499–2521.

(10)   Bicyclo[3.2.0]hept-1-ene
        298 K   39.4(0.3)   164.8(1.3)

Roth, W. R.; Adamczak, O.; Breuckman, R.; Lennartz, H.-W.; Boese, R. *Chem. Ber.* **1991**, *124*, 2499–2521.

(11) Bicyclo[3.2.0]hept-2-ene
298 K   27.8(0.1)   116.3(0.4)

Roth, W. R.; Adamczak, O.; Breuckman, R.; Lennartz, H.-W.; Boese, R. *Chem. Ber.* **1991**, *124*, 2499–2521.

(12) 4-Cyclopropylbicyclo[2.1.0]pentane
298 K   102.0(0.3)   426.8(1.3)

Roth, W. R.; Adamczak, O.; Breuckman, R.; Lennartz, H.-W.; Boese, R. *Chem. Ber.* **1991**, *124*, 2499–2521.

(13) 1,2-Dimethylenecyclopentane
298 K   55.2(0.2)   231.0(0.8)

Roth, W. R.; Adamczak, O.; Breuckman, R.; Lennartz, H.-W.; Boese, R. *Chem. Ber.* **1991**, *124*, 2499–2521.

(14) Ethylidinecyclopent-2-ene
298 K   50.5(0.1)   211.3(0.4)

Roth, W. R.; Adamczak, O.; Breuckman, R.; Lennartz, H.-W.; Boese, R. *Chem. Ber.* **1991**, *124*, 2499–2521.

(15) Bicyclo[3.2.0]hepta-1(5)-ene
298 K   40.8(0.1)   170.7(0.4)

Roth, W. R.; Adamczak, O.; Breuckman, R.; Lennartz, H.-W.; Boese, R. *Chem. Ber.* **1991**, *124*, 2499–2521.

(16) Bicyclo[4.1.0]hept-3-ene
     trans-Norcarene
298 K   92.4(0.1)   386.6(0.4)

The hydrogenation product is methylcyclohexane.

Roth, W. R.; Adamczak, O.; Breuckman, R.; Lennartz, H.-W.; Boese, R. *Chem. Ber.* **1991**, *124*, 2499–2521.

(17) Bicyclo[3.2.0]hept-6-ene
298 K   32.7(0.1)   136.8(0.4)

The product is bicyclo[3.2.0]heptane.

Roth, W. R.; Klaerner F.-G.; Lennartz, H.-W. *Chem. Ber.* **1980**, *113*, 1818–1829.

(18)   5,5-Dimethylbicyclo[2.1.0]pent-2-ene
       298 K   97.0(0.2)   405.8(0.8)

The product was dimethylcyclopentane.

Roth, W. R.; Klaerner F.-G.; Lennartz, H.-W. *Chem. Ber.* **1980**, *113*, 1818–1829.

(19)   5,5-Dimethylcyclopentadiene
       298 K   53.8(0.1)   225.1(0.4)

The product is dimethylcyclopentane.

Roth, W. R.; Klaerner F.-G.; Lennartz, H.-W. *Chem. Ber.* **1980**, *113*, 1818–1829.

(20)   Tricyclo[2.2.1.0^{2,6}]heptane
       Nortricyclane
       298 K   32.6(0.5)   136(2.1)

Hydrogenation was carried out over Rh catalyst.

Rogers, D. W.; Choi, L. S.; Girellini, R. S.; Holmes, T. J.; Allinger, N. L., *J. Phys. Chem.* **1980**, *84*, 1810–1814.

(20)   Tricyclo[2.2.1.0^{2,6}]heptane
       Nortricyclane
       Nortricyclene
       298 K   31.1(0.2)   130.1(0.8)

Hydrogenation was carried out in isooctane calorimeter fluid over Rh catalyst. The enthalpy change was corrected by 2.1 kcal mol^{-1} for 13.6% ethylcyclopentane formed during hydrogenation and by 0.1 kcal mol^{-1} for $\Delta_{vap}H$ using Kovats indices.

Flury, P.; Grob, C. A.; Wang, G. Y.; Lennatrz, H-W.; Roth, W. R. *Helvetica Chimica Acta* **1988**, *71*, 1017–1024.

(21)   *exo*-Tricyclo[3.2.0.0^{2,4}]heptane
       298 K   55.6(0.3)   232.6(1.3)

The product was bicyclo[3.2.0]heptane. The result was corrected for a small amount of methylcyclohexane in the product.

Roth, W. R.; Klaerner, F.-G.; Grimme, W.; Koeser, H. G.; Busch, R.; Muskulus, B.; Breuckmann, R.; Scholz, B. P.; Lennartz, H.-W. *Chem. Ber.* **1983**, *116*, 2717–2737.

(22) endo-Tricyclo[3.2.0.0^{2,4}]heptane
    298 K   64.1(0.3)   268.2(1.3)

The product was bicyclo[3.2.0]heptane. The result was corrected for a small amount of methylcyclohexane in the product.

Roth, W. R.; Klaerner, F.-G.; Grimme, W.; Koeser, H. G.; Busch, R.; Muskulus, B.; Breuckmann, R.; Scholz, B. P.; Lennartz, H.-W. *Chem. Ber.* **1983**, *116*, 2717–2737.

(23) Bicyclo[3.2.0.0^{2,7}]heptane
    298 K   52.6(0.2)   220.1(0.8)

Hydrogenation was carried out in isooctane calorimeter fluid over 10% Pd/C catalyst. The enthalpy change was corrected by 0.2 kcal mol^{-1} (endothermic) for the $\Delta_{vap}H$ difference between reactant and product using Kovats indices.

Flury, P.; Grob, C. A.; Wang, G. Y.; Lennatrz, H-W.; Roth, W. R. *Helvetica Chimica Acta* **1988**, *71*, 1017–1024.

(24) Tricyclo[4.1.0.0^{1,3}]heptane
    298 K   92.5(0.2)   387.0(0.8)

Roth, W. R.; Wollweber, D.; Offerhaus, R.; Rekowski, V.; Lennartz, H.-W. Sustmann, R.; Muller, W. *Chem. Ber.* **1993**, *126*, 2701–2715.

(25) 6-Methylenebicyclo[2.2.0]hexane
    298 K   88.9(0.1)

Roth, W. R.; Wollweber, D.; Offerhaus, R.; Rekowski, V.; Lennartz, H.-W. Sustmann, R.; Muller, W. *Chem. Ber.* **1993**, *126*, 2701–2715.

C_7H_{12}

(1) 1-Heptyne
    298 K   69.6(0.4)   291(1.7)

$\Delta_{hyd}H$ of all the linear monoalkynes in this series were measured in *n*-hexane solution.

Rogers, D. W.; Dagdagan, O. A.; Allinger, N. L. *J. Amer. Chem. Soc.* **1979**, *101*, 671–676.

(2) 2-Heptyne
    298 K   65.1(0.3)   272(1.3)

$\Delta_{hyd}H$ of all the linear monoalkynes in this series were measured in *n*-hexane solution.

Rogers, D. W.; Dagdagan, O. A.; Allinger, N. L. *J. Amer. Chem. Soc.* **1979**, *101*, 671–676.

(3) 3-Heptyne
     298 K   64.6(0.4)   270(1.7)

$\Delta_{hyd}H$ of all the linear monoalkynes in this series were measured in *n*-hexane solution.

Rogers, D. W.; Dagdagan, O. A.; Allinger, N. L. *J. Amer. Chem. Soc.* **1979**, *101*, 671–676.

(4) Cycloheptene
     355 K   26.5(0.2)   110.9(0.8)

Conn, J. B.; Kistiakowsky, G. B.; Smith, E. A. *J. Amer. Chem. Soc.* **1939**, *61*, 1868–1876.

(5) Cycloheptene
     298 K   25.8(0.1)   108.2(0.4)

$\Delta_{hyd}H$ was measured in acetic acid solution.

Turner, R. B.; Meador, W. R.; Winkler, R. E. *J. Amer. Chem. Soc.*. **1957**, *79*, 4116–4121.

(5') Cycloheptene
     298 K   26.4(0.1)   110.5(0.4)

Roth, W. R.; Lennartz, H.-W. *Chem. Ber.* **1980**, *113*, 1806–1817.

(6) Methylenecyclohexane
     298 K   27.8(0.1)   116.4(0.4)

$\Delta_{hyd}H$ was measured in acetic acid solution.

Turner, R. B.; Garner, R. H. *J. Amer. Chem. Soc.*. **1957**, *79*, 253 (one page).

(6') Methylenecyclohexane
     298 K   27.8(0.1)   116.1(0.4)

$\Delta_{hyd}H$ was measured in acetic acid solution.

Turner, R. B.; Garner, R. H. *J. Amer. Chem. Soc.* **1958**, *80*, 1424–1430. Steric and electronic stabilization is discussed in detail.

(7) 1-Methylcyclohexene
     298 K   25.7(0.1)   107.5(0.4)

$\Delta_{hyd}H$ was measured in acetic acid solution.

Turner, R. B.; Garner, R. H. *J. Amer. Chem. Soc.* **1957**, *79*, 253 (one page).

(7')   1-Methylcyclohexene
        298 K   25.5(0.1)   103.3(0.4)

$\Delta_{hyd}H$ was measured in acetic acid solution. Steric and electronic stabilization is discussed in detail.

Turner, R. B.; Garner, R. H. *J. Amer. Chem. Soc.* **1958**, *80*, 1424–1430.

(8)   Ethylidinecyclopentane
        298 K   24.9(0.1)   104.1(0.4)

$\Delta_{hyd}H$ was measured in acetic acid solution. Steric and electronic stabilization is discussed in detail.

Turner, R. B.; Garner, R. H. *J. Amer. Chem. Soc.* **1958**, *80*, 1424–1430.

(8')   Ethylidinecyclopentane
        298 K   24.2(0.2)   101.3(0.8)

Hydrogenation was carried out in cyclohexane.

Rogers, D. W.; Mc Lafferty, F. J. *Tetrahedron* **1971**, *27*, 3765–3775.

(8")   Ethylidinecyclopentane
        298 K   25.6(0.1)   106.9(0.4)

$\Delta_{hyd}H$ was measured in *n*-hexane solution. See also, Turner, R. B.; Garner, R. H. *J. Amer. Chem. Soc.* **1958**, *80*, 1424–1430.

Allinger, N. L.; Dodziuk, H.; Rogers, D. W.; Naik, S. N., *Tetrahedron* **1982**, *38*, 1593–1597.

(9)   1-Ethylcyclopentene
        298 K   23.5(0.1)   98.6(0.4)

$\Delta_{hyd}H$ was measured in acetic acid solution. Steric and electronic stabilization is discussed in detail.

Turner, R. B.; Garner, R. H. *J. Amer. Chem. Soc.* **1958**, *80*, 1424–1430.

(9')   1-Ethylcyclopentene
        298 K   23.5(0.2)   98.3(0.8)

Hydrogenation was carried out in cyclohexane.

Rogers, D. W.; Mc Lafferty, F. J. *Tetrahedron* **1971**, *27*, 3765–3775.

(9")   1-Ethylcyclopentene
        298 K   24.4(0.2)   101.9(0.8)

$\Delta_{hyd}H$ was measured in *n*-hexane solution.

Allinger, N. L.; Dodziuk, H.; Rogers, D. W.; Naik, S. N., *Tetrahedron* **1982**, *38*, 1593–1597.

(10)   3-Ethylcyclopentene
        298 K   27.4(0.1)   114.5(0.4)

$\Delta_{hyd}H$ was measured in *n*-hexane solution.

Allinger, N. L.; Dodziuk, H.; Rogers, D. W.; Naik, S. N., *Tetrahedron* **1982**, *38*, 1593–1597.

(11)   Vinylcyclopentane
        298 K   28.4(0.2)   118.8(0.8)

Hydrogenation was carried out in cyclohexane.

Rogers, D. W.; Mc Lafferty, F. J. *Tetrahedron* **1971**, *27*, 3765–3775.

(12)   1,2-Dimethylcyclopentene
        298 K   22.6(0.2)   94.3(0.8)

$\Delta_{hyd}H$ was measured in *n*-hexane solution. The sequence –28.5, –27, and –22.5 for $\Delta_{hyd}H$ of cyclohexene, cyclopentene, and 1,2-dimethylcyclopentene shows the influence of hydrogen atom crowding in the cyclopentane ring relative to cyclohexane (product destabilization) and reactant stabilization of the double bond in 1,2-dimethylcyclohexene by two α methyl groups.

Allinger, N. L.; Dodziuk, H.; Rogers, D. W.; Naik, S. N., *Tetrahedron* **1982**, *38*, 1593–1597.

(13)   5,5-Dimethylbicyclo[2.1.0]pentane
        298 K   55.0(0.1)   230.1(0.4)

The product was dimethylcyclopentane.

Roth, W. R.; Klaerner F.-G.; Lennartz, H.-W. *Chem. Ber.* **1980**, *113*, 1818–1829.

C_7H_{14}

(1) 1-Heptene
     355 K   30.14(0.1)   126.1(0.4)

Kistiakowsky, G. B.; Ruhoff, J. R.; Smith, H. A.; Vaughan, W. E. *J. Amer. Chem. Soc.* **1936**, *58*, 137–145.

(1') 1-Heptene
      302   29.97(0.3)   125.4(1.3)
      355 K   (gas phase, estimated)
                                        30.2(0.3)   126.6(1.3)

Williams, R. B. *J. Amer. Chem. Soc.* **1942**, *64*, 1395–1404.

Williams's measurements of $\Delta_{hyd}H$ are the first to be made using reactants in solution. His work includes a thorough discussion of the conversion of the reaction

or
$$X(\text{liquid or solution}) + H_2(\text{gas-}) \rightarrow Y(\text{in solution})$$

to
$$X(\text{liquid}) + H_2(\text{gas}) \rightarrow Y(\text{liquid})$$

$$X(\text{gas}) + H_2(\text{gas}) \rightarrow Y(\text{gas})$$

Differences in measured $\Delta_{hyd}H$ values were noted between hydrogenations carried out using Pt and Pd catalyst. No explanation was given. Pd was preferred.

(1") 1-Heptene
      298 K   30.4(0.4)   127.2(1.7)

$\Delta_{hyd}H$ carried out in *n*-hexane. Value differs slightly from the original paper due to a revised standard $\Delta_{hyd}H$, Rogers, D. W., *J. Phys. Chem.* **1979**, *83*, 2430 (one page).

Rogers, D. W.; Skanupong, S. *J. Phys. Chem.* **1974**, *78*, 256 –2572.

(1''') 1-Heptene
        298 K   30.3(0.1)   126.8(0.4)

Roth, W. R.; Lennartz, H.-W. *Chem. Ber.* **1980**, *113*, 1806–1817.

(1'''') 1-Heptene
         298 K   29.9(0.1)   125.2(0.5)

These entries are the grand mean of two experimental sets of results consisting of 9 pairs of hydrogenations (sample and standard), 18 hydrogenations in all. The uncertainties are the arithmetic mean of twice the standard deviations of each set of experimental results.

Rogers, D. W.; Dejroongruang, K. *J. Chem. Thermodynam.* **1988**, *20*, 675–680.

(2) 2-Heptene, (Z)
cis-2-Heptene
298 K   28.2(0.1)   117.9(0.4)

This result was corrected from the original value of 115.6(0.4) kJ mol^{-1} for the presence of heptane in the sample. The amount of heptane was determined by glc. These entries are the grand mean of two experimental sets of results consisting of 9 pairs of hydrogenations (sample and standard), 18 hydrogenations in all. The uncertainties are the arithmetic mean of twice the standard deviations of each set of experimental results.

Rogers, D. W.; Dejroongruang, K. *J. Chem. Thermodynam.* **1988**, *20*, 675–680.

(3) 2-Heptene, (E)
trans-2-Heptene
298 K   27.3(0.1)   114.1(0.5)

The hydrogenation was carried out in *n*-hexane. These entries are the grand mean of two experimental sets of results consisting of 9 pairs of hydrogenations (sample and standard), 18 hydrogenations in all. The uncertainties are the arithmetic mean of twice the standard deviations of each set of experimental results.

Rogers, D. W.; Dejroongruang, K. *J. Chem. Thermodynam.* **1988**, *20*, 675–680.

(4) 3-Heptene, (Z)
cis-3-Heptene
298 K   28.7(.7)   120.0(2.9)

Hydrogenation was carried out in *n*-hexane. Value differs slightly from the original paper due to a revised standard $\Delta_{hyd}H$, Rogers, D. W., *J. Phys. Chem.* **1979**, *83*, 2430 (one page).

Rogers, D. W.; Siddiqui, N. A. *J. Phys. Chem.* **1975**, *79*, 574–577.

(4') 3-Heptene, (Z)
cis-3-Heptene
298 K   28.3(0.1)   118.5(0.3)

The hydrogenation was carried out in *n*-hexane. These entries are the grand mean of two experimental sets of results consisting of 9 pairs of hydrogenations (sample and standard), 18 hydrogenations in all. The uncertainties are the arithmetic mean of twice the standard deviations of each set of experimental results.

Rogers, D. W.; Dejroongruang, K. *J. Chem. Thermodynam.* **1988**, *20*, 675–680.

(5) 3-Heptene, (E)

298 K   27.4(0.1)   114.7(0.3)

The hydrogenation was carried out in *n*-hexane. These entries are the grand mean of two experimental sets of results consisting of 9 pairs of hydrogenations (sample and standard), 18 hydrogenations in all. The uncertainties are the arithmetic mean of twice the standard deviations of each set of experimental results.

Rogers, D. W.; Dejroongruang, K. *J. Chem. Thermodynam.* **1988**, *20*, 675–680.

(6)   2-Methylhex-1-ene
      298 K   27.7(0.1)   115.8(0.4)

The hydrogenation was carried out in *n*-hexane. The methyl group in the 2- position shows substantial α stabilization relative to 1-hexene. These entries are the grand mean of two experimental sets of results consisting of 9 pairs of hydrogenations (sample and standard), 18 hydrogenations in all. The uncertainties are the arithmetic mean of twice the standard deviations of each set of experimental results.

Rogers, D. W.; Dejroongruang, K. *J. Chem. Thermodynam.* **1988**, *20*, 675–680.

(7)   3-Methylhex-1-ene
      298 K   29.8(0.1)   124.5(0.4)

The hydrogenation was carried out in *n*-hexane. Moving the methyl group from the 2-position in 2-methylhex-1-ene (previous entry) to the 3- position removes almost all of the methyl stabilization. These entries are the grand mean of two experimental sets of results consisting of 9 pairs of hydrogenations (sample and standard), 18 hydrogenations in all. The uncertainties are the arithmetic mean of twice the standard deviations of each set of experimental results.

Rogers, D. W.; Dejroongruang, K. *J. Chem. Thermodynam.* **1988**, *20*, 675–680.

(8)   4-Methylhex-1-ene
      298 K   29.3(0.1)   122.7(0.3)

The hydrogenation was carried out in *n*-hexane. Stabilization relative to 1-hexene is less than 1 kcal mol^{-1}. These entries are the grand mean of two sets of experimental results consisting of 9 pairs of hydrogenations (sample and standard), 18 hydrogenations in all. The uncertainties are the arithmetic mean of twice the standard deviations of each set of experimental results.

Rogers, D. W.; Dejroongruang, K. *J. Chem. Thermodynam.* **1988**, *20*, 675–680.

(9)   5-Methylhex-1-ene
      Neoamylethylene
      355 K   29.53(0.1)   123.6(0.4)

Hydrogenation was carried out in the gas phase. Stabilization relative to 1-hexene is less than 1 kcal mol^{-1}.

Dolliver, M. A.; Gresham, T. L.; Kistiakowsky, G. B.; Vaughan, W. E. *J. Amer. Chem. Soc.* **1937**, *59*, 831–841.

(9') 5-Methylhex-1-ene
      298 K   29.8(0.1)   124.6(0.4)

The hydrogenation was carried out in *n*-hexane. Stabilization relative to 1-hexene is less than 1 kcal mol^{-1}. These entries are the grand mean of two experimental sets of results consisting of 9 pairs of hydrogenations (sample and standard), 18 hydrogenations in all. The uncertainties are the arithmetic mean of twice the standard deviations of each set of experimental results.

Rogers, D. W.; Dejroongruang, K. *J. Chem. Thermodynam.* **1988**, *20*, 675–680.

(10) 2-Methyl-2-hexene
      298 K   26.0(0.2)   108.8(0.7)

The hydrogenation was carried out in *n*-hexane. Stabilization relative to 1-hexene is pronounced. Hydrogenation showed some kinetic lag over the normal 10-12 s instrumental response time. These entries are the grand mean of two experimental sets of results consisting of 9 pairs of hydrogenations (sample and standard), 18 hydrogenations in all. The uncertainties are the arithmetic mean of twice the standard deviations of each set of experimental results.

Rogers, D. W.; Dejroongruang, K. *J. Chem. Thermodynam.* **1988**, *20*, 675–680.

(11) 3-Methyl-2-hexene, (Z)
      3-Methyl-*cis*-2-hexene
      298 K   25.7(0.1)   107.4(0.6)

The hydrogenation was carried out in *n*-hexane. Stabilization relative to 1-hexene is pronounced. Hydrogenation showed some kinetic lag over the normal 10-12 s instrumental response time. These entries are the grand mean of two experimental sets of results consisting of 9 pairs of hydrogenations (sample and standard), 18 hydrogenations in all. The uncertainties are the arithmetic mean of twice the standard deviations of each set of experimental results.

Rogers, D. W.; Dejroongruang, K. *J. Chem. Thermodynam.* **1988**, *20*, 675–680.

(12) 3-Methylhex-2-ene, (E)
      298 K   25(?)   105(?)

The hydrogenation was carried out in *n*-hexane. This datum is unreliable. Catalyst poisoning by an impurity in the sample was suspected.

Rogers, D. W.; Dejroongruang, K. *J. Chem. Thermodynam.* **1988**, *20*, 675–680.

(13) 4-Methylhex-2-ene, (Z)
      4-Methylhex-*cis*-2-ene
      298 K   27.6(0.1)   115.6(0.3)

The hydrogenation was carried out in *n*-hexane. The result is comparable to 2-heptene, (Z). These entries are the grand mean of two experimental sets of results consisting of 9 pairs of hydrogenations (sample and standard), 18 hydrogenations in all. The uncertainties are the arithmetic mean of twice the standard deviations of each set of experimental results.

Rogers, D. W.; Dejroongruang, K. *J. Chem. Thermodynam.* **1988**, *20*, 675–680.

(14) 4-Methylhex-2-ene, (E)
      298 K   26.6(0.1)   111.2(0.4)

The hydrogenation was carried out in *n*-hexane. The result is comparable to 2-heptene, (E). These entries are the grand mean of two experimental sets of results consisting of 9 pairs of hydrogenations (sample and standard), 18 hydrogenations in all. The uncertainties are the arithmetic mean of twice the standard deviations of each set of experimental results.

Rogers, D. W.; Dejroongruang, K. *J. Chem. Thermodynam.* **1988**, *20*, 675–680.

(15) 5-Methylhex-2-ene, (Z)
      5-Methylhex-*cis*-2-ene
      298 K   27.8(0.1)   116.3(0.3)

The hydrogenation was carried out in *n*-hexane. The result is comparable to 2-heptene, (Z). These entries are the grand mean of two experimental sets of results consisting of 9 pairs of hydrogenations (sample and standard), 18 hydrogenations in all. The uncertainties are the arithmetic mean of twice the standard deviations of each set of experimental results.

Rogers, D. W.; Dejroongruang, K. *J. Chem. Thermodynam.* **1988**, *20*, 675–680.

(16) 5-Methylhex-2-ene, (E)
      298 K   26.9(0.1)   112.4(0.3)

The hydrogenation was carried out in *n*-hexane. The result is comparable to 2-heptene, (E). These entries are the grand mean of two experimental sets of results consisting of 9 pairs of hydrogenations (sample and standard), 18 hydrogenations in all. The uncertainties are the arithmetic mean of twice the standard deviations of each set of experimental results.

Rogers, D. W.; Dejroongruang, K. *J. Chem. Thermodynam.* **1988**, *20*, 675–680.

(17) 2-Methylhex-3-ene, (Z)
      2-Methylhex-*cis*-3-ene
      298 K   28.3(0.1)   118.3(0.3)

The hydrogenation was carried out in *n*-hexane. The reactant is somewhat destabilized by interference between the isopropyl and ethyl substituents in the Z configuration. These entries are the grand mean of two experimental sets of results consisting of 9 pairs of hydrogenations (sample and standard), 18 hydrogenations in all. The uncertainties are the arithmetic mean of twice the standard deviations of each set of experimental results.

Rogers, D. W.; Dejroongruang, K. *J. Chem. Thermodynam.* **1988**, *20*, 675–680.

(18) 2-Methylhex-3-ene, (E)
      298 K   27.6(0.1)   115.3(0.3)

The hydrogenation was carried out in *n*-hexane. These entries are the grand mean of two experimental sets of results consisting of 9 pairs of hydrogenations (sample and standard), 18 hydrogenations in all. The uncertainties are the arithmetic mean of twice the standard deviations of each set of experimental results.

Rogers, D. W.; Dejroongruang, K. *J. Chem. Thermodynam.* **1988**, *20*, 675–680.

(19) 3-Methylhex-3-ene, (Z)
      2-Methylhex-*cis*-3-ene
      298 K   26.4(0.1)   110.6(0.4)

The hydrogenation was carried out in *n*-hexane. Significant α methyl stabilization is observed. These entries are the grand mean of two experimental sets of results consisting of 9 pairs of hydrogenations (sample and standard), 18 hydrogenations in all. The uncertainties are the arithmetic mean of twice the standard deviations of each set of experimental results.

Rogers, D. W.; Dejroongruang, K. *J. Chem. Thermodynam.* **1988**, *20*, 675–680.

(20) 3-Methylhex-3-ene, (E)
      298 K   26.1(0.1)   109.4(0.4)

The hydrogenation was carried out in *n*-hexane. Significant α methyl stabilization is observed. These entries are the grand mean of two experimental sets of results consisting of 9 pairs of hydrogenations (sample and standard), 18 hydrogenations in all. The uncertainties are the arithmetic mean of twice the standard deviations of each set of experimental results.

Rogers, D. W.; Dejroongruang, K. *J. Chem. Thermodynam.* **1988**, *20*, 675–680.

(21) 2,3-Dimethylpent-1-ene
298 K   27.1(0.2)   113.5(0.7)

The hydrogenation was carried out in *n*-hexane. Compare with 1-pentene. Significant α methyl stabilization is observed. The entry above is the arithmetic mean of two sets of experimental results, each of 9 determinations.

Rogers, D. W.; Dejroongruang, K. *J. Chem. Thermodynam.* **1989**, *21*, 1115–1120.

(22) 2,4-Dimethylpent-1-ene
      2,4-Dimethyl-1-pentene
298 K   26.7(0.1)   111.6(0.2)

$\Delta_{hyd}H$ was measured in acetic acid. Compare with 1-pentene. Significant α methyl stabilization is observed.

Turner, R. B.; Nettleton, D. E.; Perelman, M. *J. Amer. Chem. Soc.* **1958**, *80*, 1430–1433.

(22') 2,4-Dimethylpent-1-ene
298 K   27.4(0.3)   114.6(1.2)

The hydrogenation was carried out in *n*-hexane. Compare with 1-pentene. Significant α methyl stabilization is observed. Slight kinetic lag was observed. The entry above is the arithmetic mean of two sets of experimental results, each of 9 determinations. For this compound, the estimated uncertainty was double because of kinetic lag.

Rogers, D. W.; Dejroongruang, K. *J. Chem. Thermodynam.* **1989**, *21*, 1115–1120.

(23) 3,3-Dimethylpent-1-ene
298 K   29.4(0.2)   122.9(1.0)

The hydrogenation was carried out in *n*-hexane. Compare with 1-pentene. The entry above is the arithmetic mean of two sets of experimental results, each of 9 determinations.

Rogers, D. W.; Dejroongruang, K. *J. Chem. Thermodynam.* **1989**, *21*, 1115–1120.

(24) 3,4-Dimethylpent-1-ene
298 K   29.1(0.2)   121.8(0.8)

The hydrogenation was carried out in *n*-hexane. Compare with 1-pentene. The entry above is the arithmetic mean of two sets of experimental results, each of 9 determinations.

Rogers, D. W.; Dejroongruang, K. *J. Chem. Thermodynam.* **1989**, *21*, 1115–1120.

(25) 4,4-Dimethylpent-1-ene

298 K	29.3(0.2)	122.6(0.6)

The hydrogenation was carried out in *n*-hexane. Compare with 1-pentene. The entry above is the arithmetic mean of two sets of experimental results, each of 9 determinations.

Rogers, D. W.; Dejroongruang, K. *J. Chem. Thermodynam.* **1989**, *21*, 1115–1120.

(26) 2,3-Dimethylpent-2-ene
  298 K   25.4(0.4)   106.4(1.6)

The hydrogenation was carried out in *n*-hexane. In the absence of a value for 2-pentene, compare with 2-hexene. Both stabilizing methyl groups are α to the double bond. Slight kinetic lag was observed. The entry above is the arithmetic mean of two sets of experimental results, each of 9 determinations. For this compound, the estimated uncertainty was double because of kinetic lag.

Rogers, D. W.; Dejroongruang, K. *J. Chem. Thermodynam.* **1989**, *21*, 1115–1120.

(27) 2,4-Dimethylpent-2-ene
  2,4-Dimethyl-2-pentene
  298 K   25.1(0.1)   105.2(0.4)

Typically, $\Delta_{hyd}H$ values measured in acetic acid are 0.7 – 1.0 kcal mol^{-1} less exothermic than the gas-phase values. In the absence of a value for 2-pentene, compare with 2-hexene.

Turner, R. B.; Nettleton, D. E.; Perelman, M. *J. Amer. Chem. Soc.* **1958**, *80*, 1430–1433.

(27') 2,4-Dimethylpent-2-ene
  298 K   26.1(0.3)   109.4(1.2)

The hydrogenation was carried out in *n*-hexane. In the absence of a value for 2-pentene, compare with 2-hexene. The molecule has one stabilizing methyl group in the α position. Slight kinetic lag was observed. The entry above is the arithmetic mean of two sets of experimental results, each of 9 determinations. For this compound, the estimated uncertainty was double because of kinetic lag.

Rogers, D. W.; Dejroongruang, K. *J. Chem. Thermodynam.* **1989**, *21*, 1115–1120.

(28) 3,4-Dimethylpent-2-ene, (Z)
  3,4-Dimethylpent-*cis*-2-ene
  298 K   25.9(0.1)   108.3(0.4)

The hydrogenation was carried out in *n*-hexane. In the absence of a value for 2-pentene, compare with 2-hexene. The molecule has one stabilizing methyl group in the α position.

The entry above is the arithmetic mean of two sets of experimental results, each of 9 determinations.

Rogers, D. W.; Dejroongruang, K. *J. Chem. Thermodynam.* **1989**, *21*, 1115–1120.

(29) 3,4-Dimethylpent-2-ene, (E)
      298 K   25.8(0.1)   108.0(0.4)

The hydrogenation was carried out in *n*-hexane. In the absence of a value for 2-pentene, compare with 2-hexene. The molecule has one stabilizing methyl group in the α position. The molecule has one stabilizing methyl group in the α position. The entry above is the arithmetic mean of two sets of experimental results, each of 9 determinations.

Rogers, D. W.; Dejroongruang, K. *J. Chem. Thermodynam.* **1989**, *21*, 1115–1120.

(30) 4,4-Dimethylpent-2-ene, (Z)
      4,4-Dimethyl-*cis*-2-pentene
      298 K   30.8(0.1)   128.9(0.4)

$\Delta_{hyd}H$ was measured in acetic acid. Comparison with 2-pentene indicates considerable strain destabilization of the reactant owing to opposition of the *t*-butyl group and the methyl group with which it is in a Z configuration.

Turner, R. B.; Nettleton, D. E.; Perelman, M. *J. Amer. Chem. Soc.* **1958**, *80*, 1430–1433.

(30') 4,4-Dimethylpent-2-ene, (Z)
       4,4-Dimethyl-*cis*-2-pentene
       298 K   31.2(0.1)   130.4(0.5)

The hydrogenation was carried out in *n*-hexane. Comparison with 2-pentene indicates considerable strain destabilization of the reactant owing to opposition of the *t*-butyl group and the methyl group with which it is in a Z configuration. The entry above is the arithmetic mean of two sets of experimental results, each of 9 determinations.

Rogers, D. W.; Dejroongruang, K. *J. Chem. Thermodynam.* **1989**, *21*, 1115–1120.

(31) 4,4-Dimethylpent-2-ene, (E)
      4,4-Dimethyl-*trans*-2-pentene
      298 K   26.5(0.2)   110.9(0.8)

$\Delta_{hyd}H$ was measured in acetic acid. Relaxation of the *t*-butyl group away from the methyl in the E configuration is reflected in the decrease of $\Delta_{hyd}H$ relative to 4,4-dimethylpent-2-ene, (Z) to a value that is normal for an unstrained 2-pentene (or 2-hexene).

Turner, R. B.; Nettleton, D. E.; Perelman, M. *J. Amer. Chem. Soc.* **1958**, *80*, 1430–1433.

(31')   4,4-Dimethylpent-2-ene, (E)
         298 K   27.3(0.1)   114.2(0.4)

The hydrogenation was carried out in *n*-hexane. Relaxation of the *t*-butyl group away from the methyl in the E configuration is reflected in the decrease of $\Delta_{hyd}H$ relative to dimethylpent-2-ene, (Z) to a value that is normal for an unstrained 2-pentene (or 2-hexene). The entry above is the arithmetic mean of two sets of experimental results, each of 9 determinations.

Rogers, D. W.; Dejroongruang, K. *J. Chem. Thermodynam.* **1989**, *21*, 1115–1120.

(32)   2-Ethylpent-1-ene
         298 K   27.4(0.3)   114.6(1.1)

Hydrogenation was carried out in *n*-hexane. Sample purity was only 95.8% but the error limits are relatively narrow because the major contaminants were isomers of the designated branched octene having a similar $\Delta_{hyd}H$. The entry above is the arithmetic mean of two sets of experimental results, each of 9 determinations.

Rogers, D. W.; Dejroongruang, K. *J. Chem. Thermodynam.* **1989**, *21*, 1115–1120.

(33)   3-Ethylpent-1-ene
         298 K   28.8(0.3)   120.6(1.3)

Hydrogenation was carried out in *n*-hexane. Sample purity was only 94.7% but the error limits are relatively narrow because the major contaminants were isomers of the designated branched octene having a similar $\Delta_{hyd}H$. The entry above is the arithmetic mean of two sets of experimental results, each of 9 determinations.

Rogers, D. W.; Dejroongruang, K. *J. Chem. Thermodynam.* **1989**, *21*, 1115–1120.

(34)   3-Ethylpent-2-ene
         298 K   25.5(0.2)   106.8(0.8)

Hydrogenation was carried out in *n*-hexane. Sample purity was 98.8%. The entry above is the arithmetic mean of two sets of experimental results, each of 9 determinations.

Rogers, D. W.; Dejroongruang, K. *J. Chem. Thermodynam.* **1989**, *21*, 1115–1120.

(35)   2-Ethyl-3-methylbut-1-ene
         298 K   27.2(0.3)   114.0(1.3)

Hydrogenation was carried out in *n*-hexane. Kinetic lag was observed. Sample purity was 97.1%. The entry above is the arithmetic mean of two sets of experimental results, each of 9 determinations. Kinetic effects and impurities had a relatively small effect on the error limits because kinetic lag was slight and the major contaminants were isomers of the designated branched octene having a similar $\Delta_{hyd}H$.

Rogers, D. W.; Dejroongruang, K. *J. Chem. Thermodynam.* **1989**, *21*, 1115–1120.

(36) 2,3,3-Trimethylbut-1-ene
      298 K   27.3(0.1)   114.3(0.4)

Hydrogenation was carried out in *n*-hexane. Sample purity was 99.8%. The entry above is the arithmetic mean of two sets of experimental results, each of 9 determinations.

Rogers, D. W.; Dejroongruang, K. *J. Chem. Thermodynam.* **1989**, *21*, 1115–1120.

C_8H_6

(1) Ethynylbenzene
     Phenylacetylene
     298 K   70.7(0.2)   293.7(0.8)

$\Delta_{hyd}H$ was measured in ethanol and corrected for heat of solution. The value given is for a liquid → liquid reaction corrected for $\Delta_{sol'n}H$.

Flitcroft, T. L.; Skinner, H. A. *Trans. Faraday. Soc.* **1958**, *54*, 47–53.

(2) Ethynylbenzene
     Phenylacetylene
     298 K   64.7(0.9)   270.7(3.8)

The hydrogenation was carried out in cyclohexane solution under mild conditions against a standard $\Delta_{hyd}H$ of 1-hexene so that only the exocyclic triple bond was hydrogenated.

Rogers, D. W.; Mc Lafferty, F. J. *Tetrahedron* **1971**, *27*, 3765–3775.

(3) Ethynylbenzene
     Phenylacetylene
     298 K   66.1(0.4)   276.6(0.1.7)

The hydrogenation was carried out in *n*-hexane solution under mild conditions against a standard $\Delta_{hyd}H$ of 1-hexene so that only the exocyclic triple bond was hydrogenated. The entry above is the arithmetic mean of two separate experiments consisting of 9 hydrogenation runs each.

Davis, H. E.; Allinger, N. L.; Rogers, D. W. *J. Org. Chem.* **1985**, *50*, 3601–3604.

(4) 1,5-Cyclooctadiyne
     298 K   152.9(0.3)   639.7(1.2)

The enthalpy of hydrogenation per triple bond is higher than that of acetylene due to strain induced by inclusion of the triple bonds in a C_8 ring. $\Delta_{hyd}H$ was measured in isooctane. The datum is not corrected for either $\Delta_{vap}H$ or $\Delta_{sol'n}H$.

Roth, W. R.; Hopf, H.; Horn, C. *Chem. Ber.* **1994**, *127*, 1781–1795.

C_8H_8

(1)   1,2,6,7-Octatetraene
       298 K   138.0(0.1)   577.5(0.4)

Roth, W. R.; Scholz, B. P.; Breuckmann, R.; Jelich, K.; Lennartz, H.-W. *Chem. Ber.* **1982**, *115*, 1934–1946.

(2)   Styrene
       Phenylethylene
       354   77.5(0.2)   324.2(0.8)

The product was ethylcyclohexane. The reaction was carried out in the gas phase.

Dolliver, M. A.; Gresham, T. L.; Kistiakowsky, G. B.; Vaughan, W. E. *J. Amer. Chem. Soc.* **1937**, *59*, 831–841.

(2')   Styrene
       Phenylethylene
       298 K   28.0(0.4)   117.2(1.7)

Hydrogenation was carried out under mild conditions against a standard $\Delta_{hyd}H$ of 1-hexene. Only the exocyclic double bond was hydrogenated. The sum –49.8 + (–28.0) = –77.8 kcal mol^{-1} for benzene and the exocyclic double bond in styrene can be compared to the gas-phase hydrogenation of styrene to give ethylcyclohexane, -77.5(0.2) kcal mol^{-1}.

Abboud, J.-L. M.; Jimenez, P.; Roux, V.; Turrion, C.; Lopez-Mardomingo, C.; Podosennin, A.; Rogers, D. W.; Liebman, J. F. *J. Phys. Org. Chem.* **1995**, *8*, 15–25.

(2")   Styrene
       Phenylethylene
       298 K   28.3(0.1)   118.4(0.4)

Hydrogenation was carried out under mild conditions in an electrically calibrated calorimeter. Only the exocyclic double bond was hydrogenated.

Unpublished work from the M. S. Thesis of B. Munoz-Hresko quoted in Abboud, J.-L. M.; Jimenez, P.; Roux, V.; Turrion, C.; Lopez-Mardomingo, C.; Podosennin, A.; Rogers, D. W.; Liebman, J. F. *J. Phys. Org. Chem.* **1995**, *8*, 15–25.

(3)   Cyclooctatetraene
        298 K   98.0(0.1)   409.9(0.4)

$\Delta_{hyd}H$ was measured in acetic acid. The authors give a short discussion of the history of cyclooctatetraene in resonance theory and conclude that resonance stabilization in the molecule is small. See also Turner, R. B.; Mallon, B. J.; Tichy, M.; Doering, W. von E.; Roth. W. R.; Schroeder, G. *J. Amer. Chem. Soc.* **1973**, *95*, 8605 – 8609.

Turner, D. W.; Meador, W. R.; Doering, W. von E.; Knox, L. H.; Mayer, J. R.; Wiley, D. W. *J. Amer. Chem. Soc.* **1957**, *79*, 4127–4133.

(4)   Heptafulvene
        Methylenecycloheptatriene
        298 K   92.6(0.4)   387.6(1.6)

$\Delta_{hyd}H$ was measured in diethyl carbitol. An estimated resonance energy of 13 kcal mol^{-1} is discussed.

Turner, D. W.; Meador, W. R.; Doering, W. von E.; Knox, L. H.; Mayer, J. R.; Wiley, D. W. *J. Amer. Chem. Soc.* **1957**, *79*, 4127–4133.

(5)   Benzocyclobutene
        298 K   50.3(0.1)   210.5(0.4)

$\Delta_{hyd}H$ was measured in acetic acid.

Turner, R. B.; Goebel, P.; Mallon, B. J.; von E. Doering, W.; Coburn Jr., J. F.; Pomerantz, M. *J. Amer. Chem. Soc.* **1968**, *90*, 4315–4322.

(6)   1,3-Cyclohexadiene,5,6-bis(methylene)
        298 K   54.9(0.1)   229.7(0.4)

Roth, W. and Scholz, B. P. *Chem. Ber.* **1981**, *114*, 3741–3750.

(7)   Bicyclo[2.2.2]octa-2,5,7-triene
        Barrelene
        298 K   93.8(0.3)   392.4(1.3)

Hydrogenation was carried out in acetic acid solution. $\Delta_{hyd}H$ of the first double bond to form bicyclo[2.2.2]octa-2,5-diene was –37.6 kcal mol^{-1} as determined by difference (see Turner, R. B.; Meador, W. R.; Winkler, R. E. *J. Amer. Chem. Soc..* **1957**, *79*, 4116–4121) reflecting a high level of strain in the reactant molecule and little or no delocalization stabilization in the diene.

Turner, R. B. *J. Am. Chem. Soc.*, **1964**, *86*, 3586–3587.

C_8H_{10}

(1) 1,7-Octadiyne
298 K   139.7(1.2)   584.5(5.0)

$\Delta_{hyd}H$ was measured in glacial acetic acid and corrected for heat of solution. The value given is for a liquid → liquid reaction.

Flitcroft, T. L,; Skinner, H. A.; Whiting, M. C. *Trans. Faraday Soc.* **1957**, *53*, 784–790.

(2) 1,2-Divinylcyclobutene
298 K   81.4(0.1)   340.6(0.4)

Roth, W. R.; Scholz, B. P.; Breuckmann, R.; Jelich, K.; Lennartz, H.-W. *Chem. Ber.* **1982**, *115*, 1934–1946.

(3) Ethylbenzene
354   48.9(0.1)   204.7(0.4)

Dolliver, M. A.; Gresham, T. L.; Kistiakowsky, G. B.; Vaughan, W. E. *J. Amer. Chem. Soc.* **1937**, *59*, 831–841.

(4) o-Xylene
355 K   47.2(0.2)   197.7(0.8)

$\Delta_{hyd}H$ was measured in the gas phase.

Dolliver, M. A.; Gresham, T. L.; Kistiakowsky, G. B.; Vaughan, W. E. *J. Amer. Chem. Soc.* **1937**, *59*, 831–841.

(5) Bicyclo[2.2.2]octadiene
298 K   56.2(0.1)   235.1(0.4)

$\Delta_{hyd}H$ was measured in acetic acid solution.

Turner, R. B.; Meador, W. R.; Winkler, R. E. *J. Amer. Chem. Soc..* **1957**, *79*, 4116–4121.

(6) 1,3,5-Cyclooctatriene
298 K   72.4(0.3)   302.8(1.2)

$\Delta_{hyd}H$ was measured in acetic acid. See also the discussion on cyclooctetraene.

Turner, D. W.; Meador, W. R.; Doering, W. von E.; Knox, L. H.; Mayer, J. R.; Wiley, D. W. *J. Amer. Chem. Soc.* **1957**, *79*, 4127–4133.

(6') 1,3,5-Cyclooctatriene
298 K   76.4(0.4)   319.7(1.2)

The hydrogenation was carried out in acetic acid solution. This value supersedes the previous one (6). See also the discussion on cyclooctatetraene in Turner, D. W.; Meador, W. R.; Doering, W. von E.; Knox, L. H.; Mayer, J. R.; Wiley, D. W. *J. Amer. Chem. Soc.* **1957**, *79*, 4127–4133.

Turner, R. B.; Mallon, B. J.; Tichy, M.; Doering, W. von E.; Roth. W. R.; Schroeder, G. *J. Amer. Chem. Soc.* **1973**, *95*, 8605–8609.

(7)   1,3,6-Cyclooctatriene
       298 K   79.9(0.2)   334.3(0.8)

The hydrogenation was carried out in acetic acid solution.

Turner, R. B.; Mallon, B. J.; Tichy, M.; Doering, W. von E.; Roth. W. R.; Schroeder, G. *J. Amer. Chem. Soc.* **1973**, *95*, 8605–8609.

(8)   Bicyclo[4.2.0]octa-2,4-diene
       298 K   51.6(0.1)   216.0(0.4)

$\Delta_{hyd}H$ was measured in acetic acid.

Turner, R. B.; Goebel, P.; Mallon, B. J.; von E. Doering, W.; Coburn Jr., J. F.; Pomerantz, M. *J. Amer. Chem. Soc.* **1968**, *90*, 4315–4322.

(9)   Bicyclo[4.2.0]octa-1,5-diene
       298 K   59.5(0.1)   248.9(0.4)

Roth, W. R.; Scholz, B. P.; Breuckmann, R.; Jelich, K.; Lennartz, H.-W. *Chem. Ber.* **1982**, *115*, 1934–1946.

(10)   Bicyclo[3.2.1]octa-2,6-diene
       298 K   57.9(0.3)   242.1(1.2)

$\Delta_{hyd}H$ was measured in acetic acid solution. Solvent effects are significant for the dienes and may cause the measured enthalpy changes to be as much as 1.5 kcal mol^{-1} less exothermic than the gas-phase values.

Turner, R. B.; Jarrett, A. D.; Goebel, P.; Mallon, B. J. *J. Amer. Chem. Soc.* **1973**, *95*, 790–792.

(11)   *iso*-Propylidinecyclopenta-1,3-diene
       298 K   72.2(0.1)   302.1(0.4)

Roth, W. R.; Adamczak, O.; Breuckman, R.; Lennartz, H.-W.; Boese, R. *Chem. Ber.* **1991**, *124*, 2499–2521.

(12)   Bicyclo[3.2.1]octa-2,6-diene

298 K   60.8(0.2)   254.4(0.8)

Roth, W. R.; Adamczak, O.; Breuckman, R.; Lennartz, H.-W.; Boese, R. *Chem. Ber.*
1991, *124*, 2499–2521.

(13)   *exo*-Tricyclo[3.2.1.0^{2,4}]oct-6-ene
        298 K   74.7(0.1)   312.5(0.4)

Both the *exo*- and *endo*- forms involve hydrogenation of a double bond and hydrogenation of a cyclopropane ring which sometimes behaves like a double bond. Both compounds are essentially norbornadiene with a cyclopropane ring replacing one of the double bonds in either the *exo*- position or the *endo*- position. They hydrogenate mainly to methylnorbornane where the methyl group resulting from the rupture of the cyclopentane ring is in either the *exo*- or *endo*- position according to the reactant conformation.

Roth, W. R.; Adamczak, O.; Breuckman, R.; Lennartz, H.-W.; Boese, R. *Chem. Ber.*
1991, *124*, 2499–2521.

(14)   *endo*-Tricyclo[3.2.1.0^{2,4}]oct-6-ene
        298 K   76.0(0.2)   318.0(0.8)

Both the *exo*- and *endo*- forms involve hydrogenation of a double bond and hydrogenation of a cyclopropane ring, which sometimes behaves like a double bond. Both compounds are essentially norbornadiene with a cyclopropane ring replacing one of the double bonds in either the *exo*- position or the *endo*- position. They hydrogenate mainly to methylnorbornane where the methyl group resulting from the rupture of the cyclopentane ring is in either the *exo*- or *endo*- position according to the reactant conformation.

Roth, W. R.; Adamczak, O.; Breuckman, R.; Lennartz, H.-W.; Boese, R. *Chem. Ber.*
1991, *124*, 2499–2521.

(15)   Bishomoprismane
        298 K   66.1   276.6

The catalyst was 5% Pt/C. No experimental uncertainty is given because only one measurement was made. A mixture of products was obtained. The enthalpy of formation was calculated for bicyclo[3.2.1]octane.

Bishomoprismane is a structural isomer of the tricyclo[3.2.1.0^{2,4}]oct-6-enes. It resembles prismane except that two of the lateral edges of the prism have interposed methylene groups. The *exo*- conformer of tricyclo[3.2.1.0^{2,4}]oct-6-ene is readily converted into bishomoprismane (Rh catalyst) but the *endo*- form is not.

Roth, W. R.; Adamczak, O.; Breuckman, R.; Lennartz, H.-W.; Boese, R. *Chem. Ber.*
1991, *124*, 2499–2521.

(16)   Bicyclo[3.3.0]octa-2,7-diene
        Biquinacene
        298 K   54.2(0.5)   227(2)

The entry above is the arithmetic mean of three sets of experimental results, each of 9 determinations.

Rogers, D. W.; Loggins, S. A.; Samuel, S. D.; Finnerty, M. A.; Liebman, J. F. *Structural Chemistry* **1990**, *1*, 481–489.

(17)   Bicyclo[2.2.0]hexane, 2,3-bis(methylene)
        298 K   116.8(0.1)   488.7(0.4)

Roth, W. R.; Scholz, B. P.; Breuckmann, R.; Jelich, K.; Lennartz, H.-W. *Chem. Ber.* **1982**, *115*, 1934–1946.

(18)   2,3-divinylbutadiene
        298 K   112.7(0.2)   471.5(0.8)

Roth, W. R.; Scholz, B. P.; Breuckmann, R.; Jelich, K.; Lennartz, H.-W. *Chem. Ber.* **1982**, *115*, 1934–1946.

(19)   7-methylenebicyclo[3.2.0]hept-1-ene
        298 K   68.3(0.1)   285.8(0.4)

The hydrogenation was carried out in isooctane and corrected for enthalpies of solution, not however for the heat of vaporization.

Roth, W. R.; Ruhkamp, J; Lennartz, H.-W. *Chem. Ber.* **1991**, *124*, 2047–2051.

(20)   1-Methylenespiro[2.4]hept-4-ene
        298 K   101.1(0.1)   423.0(0.4)

The hydrogenation was carried out in isooctane and corrected for enthalpies of solution, not however for the heat of vaporization. There were three reaction products, propylcyclohexane, isopropylcyclohexane, and methylethylcyclohexane. The error estimate is semiquantitative because the analytical determination of the three reaction products contribute to the error and the enthalpies of formation of the products were calculated by MM2ERW force field calculations.

Roth, W. R.; Winzer, M.; Korell, M; Wildt, H. *Liebigs Ann.* . **1995**, 897–919.

C_8H_{12}

(1)   3-Octyne-1-ene
        298 K   93.4(1.5)   390.8(6.3)

The value given is for a liquid → liquid reaction corrected for $\Delta_{sol'n} H$. Separate measurements were made in ethanol and acetic acid. Hydrogen uptake was abnormal indicating the possibility of partial polymerization

Flitcroft, T. L.; Skinner, H. A. *Trans. Faraday Soc.* **1958**, *54*, 47–53.

(2)   1,3-Cyclooctadiene, (Z,Z)
       cis-cis-1,3-Cyclooctadiene
       298 K   49.0(0.1)   204.8(0.4)

The hydrogenation was carried out in acetic acid solution.

Turner, R. B.; Mallon, B. J.; Tichy, M.; Doering, W. von E.; Roth. W. R.; Schroeder, G. *J. Amer. Chem. Soc.* **1973**, *95*, 8605–8609.

(2')   1,3-Cyclooctadiene, (Z,Z)
       cis,cis-1,3-Cyclooctadiene
       298 K   49.8   208.4

One measurement was made, hence no experimental uncertainty was given.

Roth, W. R.; Adamczak, O.; Breuckman, R.; Lennartz, H.-W.; Boese, R. *Chem. Ber.* **1991**, *124*, 2499–2521.

(3)   1,3-Cyclooctadiene, (Z,E)
       cis,trans-1,3-Cyclooctadiene
       298 K   64.7(0.1)   270.7(0.4)

The difference between the Z,Z isomer and the Z,E isomer (15 kcal mol^{-1}) is quite dramatic. Compare the Z and E isomers of cyclooctene, Rogers, D. W.; von Voithenberg, H.; Allinger, N. L. *J. Org. Chem.* **1978**, *43*, 360–361 (11 kcal mol^{-1}).

Roth, W. R.; Adamczak, O.; Breuckman, R.; Lennartz, H.-W.; Boese, R. *Chem. Ber.* **1991**, *124*, 2499–2521.

(4)   1,4-Cyclooctadiene, (Z,Z)
       cis,cis-1,4-Cyclooctadiene
       298 K   52.1(0.3)   217.9(1.2)

The hydrogenation was carried out in acetic acid solution.

Turner, R. B.; Mallon, B. J.; Tichy, M.; Doering, W. von E.; Roth. W. R.; Schroeder, G. *J. Amer. Chem. Soc.* **1973**, *95*, 8605–8609.

(5)   1,5-Cyclooctadiene, (Z,Z)
       cis,cis-1,5-Cyclooctadiene

298 K 53.7(0.1) 224.6(0.4)

The hydrogenation was carried out in acetic acid solution.

Turner, R. B.; Mallon, B. J.; Tichy, M.; Doering, W. von E.; Roth. W. R.; Schroeder, G. *J. Amer. Chem. Soc.* **1973**, *95*, 8605–8609.

(5') 1,5-Cyclooctadiene, (Z,Z)
cis, cis-1,5-Cyclooctadiene
298 K   55.0(0.2)   230.1(0.8)

$\Delta_{hyd}H$ was measured in an inert solvent and corrected for the (small) enthalpy of solution.

Roth, W. R.; Adamczak, O.; Breuckman, R.; Lennartz, H.-W.; Boese, R. *Chem. Ber.* **1991**, *124*, 2499–2521.

(6) 1,5-Cyclooctadiene, (Z,E)
cis, trans-1,5-Cyclooctadiene
298 K   67.5(0.1)   282.4(0.4)

$\Delta_{hyd}H$ was measured in an inert solvent and corrected for the (small) enthalpy of solution.

Roth, W. R.; Adamczak, O.; Breuckman, R.; Lennartz, H.-W.; Boese, R. *Chem. Ber.* **1991**, *124*, 2499–2521.

(7) 1,5-Cyclooctadiene, (E,E)
trans, trans-1,5-Cyclooctadiene
298 K   76.6(1.0)   320.5(4.1)

Roth, W. R.; Adamczak, O.; Breuckman, R.; Lennartz, H.-W.; Boese, R. *Chem. Ber.* **1991**, *124*, 2499–2521.

(8) Bicyclo[2.2.2]octene
298 K   28.2(0.2)   118.0(0.8)

$\Delta_{hyd}H$ was measured in acetic acid solution.

Turner, R. B.; Meador, W. R.; Winkler, R. E. *J. Amer. Chem. Soc..* **1957**, *79*, 4116–4121.

(9) Cyclooctyne
298 K   69.0 --   289--

Incomplete reaction was suspected. This is a lower limit. $\Delta_{hyd}H$ was measured in acetic acid solution. Solvent effects are significant for the alkynes and may cause the measured enthalpy changes to be as much as 2.5 kcal mol^{-1} less exothermic than the gas-phase values.

Turner, R. B.; Jarrett, A. D.; Goebel, P.; Mallon, B. J. *J. Amer. Chem. Soc.* **1973**, *95*, 790–792.

(9')   Cyclooctyne
       298 K   69.6(0.2)   291.2(0.8)

Roth, W. R.; Hopf, H.; Horn, C. *Chem. Ber.* **1994**, *127*, 1781–1795. $\Delta_{hyd}H$ was measured in isooctane. The datum is not corrected for either $\Delta_{vap}H$ or $\Delta_{sol'n}H$.

(10)   Bicyclo[3.2.1]oct-2-ene
       298 K   26.8(0.1)   112.3(0.4)

$\Delta_{hyd}H$ was measured in acetic acid solution. Solvent effects are significant for the alkenes and may cause the measured enthalpy changes to be as much as 0.7 kcal mol^{-1} less exothermic than the corrected values.

Turner, R. B.; Jarrett, A. D.; Goebel, P.; Mallon, B. J. *J. Amer. Chem. Soc.* **1973**, *95*, 790–792.

(11)   *iso*-Propylidinecyclopent-2-ene
       298 K   49.7(0.4)   207.9(1.6)

Roth, W. R.; Adamczak, O.; Breuckman, R.; Lennartz, H.-W.; Boese, R. *Chem. Ber.* **1991**, *124*, 2499–2521.

(12)   Bicyclo[4.2.0]oct-1(4)ene
       298 K   29.4(0.1)   123.0(0.4)

The product of hydrogenation is bicyclo[4.2.0]octane.

Roth, W. R.; Adamczak, O.; Breuckman, R.; Lennartz, H.-W.; Boese, R. *Chem. Ber.* **1991**, *124*, 2499–2521.

(13)   2,5-Dimethylhexa-1,3,5-triene
       298 K   76.1(0.1)   318.4(0.4)

$\Delta_{hyd}H$ was measured in an inert solvent and corrected for the (small) enthalpy of solution.

Roth, W. R.; Adamczak, O.; Breuckman, R.; Lennartz, H.-W.; Boese, R. *Chem. Ber.* **1991**, *124*, 2499–2521.

(14)   1,2-Divinylcyclobutane, (E)

trans-1,2-Divinylcyclobutane
298 K   62.2(0.1)   260.2(0.4)

$\Delta_{hyd}H$ was calculated for *trans*-1,2-diethylcyclobutane as the reaction product. The authors note a serious discrepancy (~ 8 kcal mol^{-1} = 33 kJ mol^{-1}) with the literature value for $\Delta_f H$ calculated from a measurement of the energy of combustion.

Roth, W. R.; Adamczak, O.; Breuckman, R.; Lennartz, H.-W.; Boese, R. *Chem. Ber.* **1991**, *124*, 2499–2521.

(15)   3-Vinyl,4-methylpenta-1,3-diene
       298 K   81.3(0.1)   340.2(0.4)

Roth, W. R.; Adamczak, O.; Breuckman, R.; Lennartz, H.-W.; Boese, R. *Chem. Ber.* **1991**, *124*, 2499–2521.

(16)   Bicyclo[4.2.0]oct-7-ene
       298 K   35.2(0.1)   147.3(0.4)

The product is bicyclo[4.2.0]octane. $\Delta_{hyd}H$ was measured in an inert solvent and corrected for the (small) enthalpy of solution.

Roth, W. R.; Klaerner F.-G.; Lennartz, H.-W. *Chem. Ber.* **1980**, *113*, 1818–1829.

(17)   Bicyclo[3.3.0]oct-2-ene
       Dihydrobiquinacene
       298 K   26.8(0.3)   112(1)

The entry above is the arithmetic mean of three sets of experimental results, each of 9 determinations.

Rogers, D. W.; Loggins, S. A.; Samuel, S. D.; Finnerty, M. A.; Liebman, J. F. *Structural Chemistry* **1990**, *1*, 481–489.

(18)   Tricyclo[3.2.1.0^{1,5}]octane
       298 K   71.6(0.3)   299.6(1.2)

Both the bridge and the cyclopropane ring are broken on hydrogenation. The enthalpy of formation given (see original paper) was calculated on the basis of methylcycloheptane as the reaction product.

Roth, W. R.; Adamczak, O.; Breuckman, R.; Lennartz, H.-W.; Boese, R. *Chem. Ber.* **1991**, *124*, 2499–2521.

C_8H_{14}

(1) 1-Octyne
     298 K   69.2(0.6)   290(2.5)

The difference between 1-octyne and 2-octyne is a typical "terminal effect" for alkynes. $\Delta_{hyd}H$ of all the linear monoalkynes in this series were measured in *n*-hexane solution.

Rogers, D. W.; Dagdagan, O. A.; Allinger, N. L. *J. Amer. Chem. Soc.* **1979**, *101*, 671–676.

(2) 2-Octyne
     298 K   65.1(0.1)   272.4(0.4)

$\Delta_{hyd}H$ of all the linear monoalkynes in this series were measured in *n*-hexane solution. The slight trend toward less exothermic enthalpies of hydrogenation for 2-octyne to 4-octyne is due to the larger number of conformers in the alkane product. Thus, $\Delta_f H$ of the alkyne is not *decreasing* along this series, $\Delta_f H$ of the conformational mix of products is *increasing*. Compare the linear nonynes and decynes.

Rogers, D. W.; Dagdagan, O. A.; Allinger, N. L. *J. Amer. Chem. Soc.* **1979**, *101*, 671–676.

(3) 3-Octyne
     298 K   64.8(0.2)   271.1(0.8)

Rogers, D. W.; Dagdagan, O. A.; Allinger, N. L. *J. Amer. Chem. Soc.* **1979**, *101*, 671–676.

$\Delta_{hyd}H$ of all the linear monoalkynes in this series were measured in *n*-hexane solution.

(4) 4-Octyne
     298 K   64.2(0.3)   269(1.2)

Rogers, D. W.; Dagdagan, O. A.; Allinger, N. L. *J. Amer. Chem. Soc.* **1979**, *101*, 671–676.

(4') 4-Octyne
      298 K   62.8(0.2)   262.8(0.8)

$\Delta_{hyd}H$ was measured in acetic acid solution. Solvent effects are significant for the alkynes and may cause the measured enthalpy changes to be as much as 2.5 kcal mol^{-1} less exothermic than values measured in the gas phase or in an inert solvent. See also Sicher, J.; Svoboda, M.; Zavada, J.; Turner, R. B.; Goebel, P. *Tetrahedron*, **1966**, *22*, 659–671.

Turner, R. B.; Jarrett, A. D.; Goebel, P.; Mallon, B. J. *J. Amer. Chem. Soc.* **1973**, *95*, 790–792.

(5)   Cyclooctene, (Z)
        cis-Cyclooctene
        355 K   23.5(0.2)   98.4(0.8)

Conn, J. B.; Kistiakowsky, G. B.; Smith, E. A. *J. Amer. Chem. Soc.* **1939**, *61*, 1868–1876.

(5')   Cyclooctene, (Z)
        cis-Cyclooctene
        298 K   23.0(0.1)   96.2(0.4)

$\Delta_{hyd}H$ was measured in acetic acid solution.

Turner, R. B.; Meador, W. R. *J. Amer. Chem. Soc..* **1957**, *79*, 4133–4136. See also Turner, R. B.; Mallon, B. J.; Tichy, M.; Doering, W. von E.; Roth. W. R.; Schroeder, G. *J. Amer. Chem. Soc.* **1973**, *95*, 8605–8609.

(5")   Cyclooctene, (Z)
        cis-Cyclooctene
        298 K   22.4(0.3)   93.7(1.3)

The hydrogenation was carried out in glacial acetic acid and corrected for solvent effects. The reaction rate was somewhat sluggish.

Rogers, D. W.; Mc Lafferty, F. J. *Tetrahedron* **1971**, *27*, 3765–3775.

(5''') Cyclooctene, (Z)
        cis-Cyclooctene
        298 K   23.0(0.2)   96.4(0.8)

$\Delta_{hyd}H$ was measured in an inert solvent. The enthalpy of solution was assumed to be negligible.

Rogers, D. W.; von Voithenberg, H.; Allinger, N. L. *J. Org. Chem.* **1978**, *43*, 360–361.

(5'''') Cyclooctene, (Z)
        cis-Cyclooctene
        298 K   24.5(0.2)   102.5(0.8)

Roth, W. R.; Lennartz, H.-W. *Chem. Ber.* **1980**, *113*, 1806–1817.

(5''''') Cyclooctene, (Z)
        cis-Cyclooctene
        298 K   24.3(0.2)   101.7(0.8)

von E. Doering, W.; Roth, W. R.; Bauer, F.; Boenke, M.; Breuckmann, R.; Ebbrecht, M; Herbold, M.; Schmidt, R.; Lennartz, H.-W.; Lenoir, D.; Boese, R. *Chem. Ber.* **1989**, *122*, 1263–1275.

(6) Cyclooctene, (E)
     trans-Cyclooctene
     298 K   32.2(0.2)   134.9(0.8)

$\Delta_{hyd}H$ was measured in acetic acid solution. Boiling points of the *cis* and *trans* isomers are nearly identical but the two $\Delta_{hyd}H$ values differ by about 10 kcal mol^{-1}, suggesting possible contamination of the reactant. See the next entry.

Turner, R. B.; Meador, W. R. *J. Amer. Chem. Soc..* **1957**, *79*, 4133–4136. See also Turner, R. B.; Mallon, B. J.; Tichy, M.; Doering, W. von E.; Roth. W. R.; Schroeder, G. *J. Amer. Chem. Soc.* **1973**, *95*, 8605–8609.

(6') Cyclooctene, (E)
      trans-Cyclooctene
      298 K   34.4(0.4)   143.9(1.6)

The reactant is not easy to purify. $\Delta_{hyd}H$ was measured in an inert solvent. Compare cyclooctadiene Roth, W. R.; Adamczak, O.; Breuckman, R.; Lennartz, H.-W.; Boese, R. *Chem. Ber.* **1991**, *124*, 2499–2521.

Rogers, D. W.; von Voithenberg, H.; Allinger, N. L. *J. Org. Chem.* **1978**, *43*, 360–361.

(6") Cyclooctene, (E)
      trans-cyclooctene
      298 K   34.5(0.1)   148.5(0.4)

Roth, W. R.; Adamczak, O.; Breuckman, R.; Lennartz, H.-W.; Boese, R. *Chem. Ber.* **1991**, *124*, 2499–2521.

(6''') Cyclooctene, (E)
        trans-cyclooctene
        298 K   34.1(0.3)   142.7(1.2)

von E. Doering, W.; Roth, W. R.; Bauer, F.; Boenke, M.; Breuckmann, R.; Ebbrecht, M; Herbold, M.; Schmidt, R.; Lennartz, H.-W.; Lenoir, D.; Boese, R. *Chem. Ber.* **1989**, *122*, 1263–1275.

(7) 3,3-dimethylcyclohexene
     298 K   27.7(0.2)   115.7(0.8)

$\Delta_{hyd}H$ was measured in acetic acid solution. The compound is slightly stabilized relative to cyclohexene. Destabilization of the reaction product is also possible.

Turner, R. B.; Jarrett, A. D.; Goebel, P.; Mallon, B. J. *J. Amer. Chem. Soc.* **1973**, *95*, 790–792.

(8) 4,4-dimethylcyclohexene
298 K   26.6(0.1)   111.4(0.4)

$\Delta_{hyd}H$ was measured in acetic acid solution. Stabilization relative to cyclohexene is nearly 2 kcal mol^{-1}. Destabilization of the reaction product is also possible.

Turner, R. B.; Jarrett, A. D.; Goebel, P.; Mallon, B. J. 197*J. Amer. Chem. Soc.* **1973**, *95*, 790–792.

(9) Methylenecycloheptane
298 K   26.3(0.1)   109.9(0.4)

$\Delta_{hyd}H$ was measured in acetic acid solution.

Turner, R. B.; Garner, R. H. *J. Amer. Chem. Soc.* **1958**, *80*, 1424–1430.

(10) 1-Methylcycloheptene
298 K   24.0(0.1)   100.5(0.4)

$\Delta_{hyd}H$ was measured in acetic acid solution. The observation that 1-methylcycloheptene is more stable than methylenecycloheptane continues the trend found for the corresponding 5- and 6-membered rings, Turner, R. B.; Garner, R. H. *J. Amer. Chem. Soc.* **1957**, *79*, 253 (one page).

Turner, R. B.; Garner, R. H. *J. Amer. Chem. Soc.* **1958**, *80*, 1424–1430.

(11) Ethylidinecyclohexane
298 K   26.3(0.1)   110.1(0.4)

$\Delta_{hyd}H$ was measured in acetic acid solution.

Turner, R. B.; Garner, R. H. *J. Amer. Chem. Soc.* **1958**, *80*, 1424–1430.

(11') Ethylidinecyclohexane
298 K   26.2(0.3)   109.6(1.2)

The hydrogenation was carried out in cyclohexane.

Rogers, D. W.; Mc Lafferty, F. J. *Tetrahedron* **1971**, *27*, 3765–3775.

(12) 1-Ethylcyclohexene
298 K   25.1(0.1)   105.0(0.4)

$\Delta_{hyd}H$ was measured in acetic acid solution. The observation that 1-methylcycloheptene is more stable than methylenecycloheptane continues the trend found for the corresponding 5- and 6-membered rings, Turner, R. B.; Garner, R. H. *J. Amer. Chem. Soc.* **1957**, *79*, 253 (one page).

Turner, R. B.; Garner, R. H. *J. Amer. Chem. Soc.* **1958**, *80*, 1424–1430.

(13) Propenylcyclopentane
      Allylcyclopentane
      298 K   30.9(0.4)   129.3(1.6)

The hydrogenation was carried out in cyclohexane.

Rogers, D. W.; Mc Lafferty, F. J. *Tetrahedron* **1971**, *27*, 3765–3775.

(14) Cyclohexylethene
      Vinylcyclohexane
      298 K   27.9(0.8)   116.7(3.3)

The hydrogenation was carried out in cyclohexane.

Rogers, D. W.; Mc Lafferty, F. J. *Tetrahedron* **1971**, *27*, 3765–3775.

(15) 3,4-Dimethylhexa-2,4-diene, (E,E)
      298 K   50.6(0.2)   211.7(0.8)

The hydrogenation was carried out in isooctane and corrected for the double bond increment in the heat of solution and for estimated differences in $\Delta_{vap}H$ (see original publication, Table IV).

Roth, W. R.; Lennartz, H.-W.; von E. Doering; Dolbier Jr., W. R.; Schmidhauser, J. C. *J. Amer. Chem. Soc.* **1988**, *110*, 1883–1889.

(16) 3,4-Dimethylhexa-2,4-diene, (Z,Z)
      298 K   50.2(0.2)   210.1(0.8)

The hydrogenation was carried out in isooctane and corrected for the double bond increment in the heat of solution and for estimated differences in $\Delta_{vap}H$ (see original publication, Table IV).

Roth, W. R.; Lennartz, H.-W.; von E. Doering; Dolbier Jr., W. R.; Schmidhauser, J. C. *J. Amer. Chem. Soc.* **1988**, *110*, 1883–1889.

(17) 3,4-Dimethylhexa-2,4-diene, (E,Z)
      298 K   51.9(0.2)   217.1(0.8)

The hydrogenation was carried out in isooctane and corrected for the double bond increment in heat of solution and for estimated differences in $\Delta_{vap}H$ (see original publication, Table IV).

Roth, W. R.; Lennartz, H.-W.; von E. Doering; Dolbier Jr., W. R.; Schmidhauser, J. C. *J. Amer. Chem. Soc.* **1988**, *110*, 1883–1889.

C_8H_{16}

(1) 1-Octene
     299.1   30.1(0.3)   125.8(1.2)

Hydrogenation was carried out in an oscillatory calorimeter using glacial acetic acid as the calorimeter fluid. Results are corrected for $\Delta_{sol'n}H$ of the product in acetic acid.

Bretschneider, E.; Rogers, D. W. *Mikrochem. Acta* [Wien] **1970**, 482–490.

(1') 1-Octene
      298 K   28.3(0.6)   118.4(2.5)

This result was measured in an early version of a new calorimeter design and is not preferred. Hydrogenation was carried out in cyclohexane. An error making the measured value about 1.7 kcal mol^{-1} less exothermic than the true value is likely.

Rogers, D. W.; Mc Lafferty, F. J. *Tetrahedron* **1971**, *27*, 3765–3775.

(1") 1-Octene
      298 K   30.1(0.3)   125.9(1.2)

Hydrogenation was carried out in *n*-hexane. The value differs slightly from the original paper due to a revised standard $\Delta_{hyd}H$, Rogers, D. W., *J. Phys. Chem.* **1979**, *83*, 2430 (one page).

Rogers, D. W.; Skanupong, S. *J. Phys. Chem.* **1974**, *78*, 2569–2572.

(1''') 1-Octene
        298 K   30.0(0.3)   125.6(1.2)

This paper is part of a series in which the concept of a "thermoneutral solvent as the calorimeter fluid is carried to its most advantageous level". If a straight chain octene is hydrogenated using octane as the calorimeter fluid, the reaction product is *n*-octane and the enthalpy of solution of the product in the calorimeter fluid is necessarily zero, causing no error. A detailed discussion of methodological error is given along with a limiting value of about 0.2 kcal mol^{-1} = 0.8 kJ mol^{-1}. In this work, methodological error has been reduced to a level that is below the error due to sample impurities.

Rogers, D. W.; Dejroongruang, K.; Samuel, S. D.; Fang, W.; Zhao, Y. *J. Chem. Thermodynam.* **1992**, *24*, 561–565.

(2) 2-Octene, (Z)
    cis-2-Octene
    298 K   28.5(0.3)   119.4(1.1)

The value given is the arithmetic mean of two experiments. Uncertainties are the arithmetic mean of the uncertainties for each experiment which, in turn, are twice the standard deviation from the mean for four degrees of freedom.

Rogers, D. W.; Dejroongruang, K.; Samuel, S. D.; Fang, W.; Zhao, Y. *J. Chem. Thermodynam.* **1992**, *24*, 561–565.

(3) 2-Octene, (E)
    298 K   27.6(0.2)   115.5(0.7)

The "end effect" of ~ 2.5 kcal mol^{-1} is seen by comparing 2-octene, (E) with 1-octene. The E-Z energy difference of 0.9 kcal mol^{-1} found by Kistiakowsky for the 2-butenes is also found here by comparing the E and Z isomer pairs among the octenes.

Rogers, D. W.; Dejroongruang, K.; Samuel, S. D.; Fang, W.; Zhao, Y. *J. Chem. Thermodynam.* **1992**, *24*, 561–565.

(4) 3-Octene, (Z)
    cis-3-Octene
    298 K   28.2(0.1)   117.8(0.4)

The value given is the arithmetic mean of two experiments. Uncertainties are the arithmetic mean of the uncertainties for each experiment which, in turn, are twice the standard deviation from the mean for four degrees of freedom.

Rogers, D. W.; Dejroongruang, K.; Samuel, S. D.; Fang, W.; Zhao, Y. *J. Chem. Thermodynam.* **1992**, *24*, 561–565.

(5) 3-Octene, (E)
    298 K   27.7(0.1)   115.8(0.4)

The value given is the arithmetic mean of two experiments. Uncertainties are the arithmetic mean of the uncertainties for each experiment which, in turn, are twice the standard deviation from the mean for four degrees of freedom.

Rogers, D. W.; Dejroongruang, K.; Samuel, S. D.; Fang, W.; Zhao, Y. *J. Chem. Thermodynam.* **1992**, *24*, 561–565.

(6) 4-Octene, (Z)
    cis-4-Octene

| | 298 K | 27.4(0.1) | 114.6(0.4) |

$\Delta_{hyd}H$ was measured in acetic acid solution.

Turner, R. B.; Jarrett, A. D.; Goebel, P.; Mallon, B. J. *J. Amer. Chem. Soc.* **1973**, *95*, 790–792.

(6') 4-Octene, (Z)
     cis-4-Octene
     298 K   29.3(0.5)   122.6(2.1)

The hydrogenation was carried out in *n*-hexane. The value differs slightly from the original paper due to a revised standard $\Delta_{hyd}H$, Rogers, D. W., *J. Phys. Chem.* **1979**, *83*, 2430 (one page).

Rogers, D. W.; Siddiqui, N. A. *J. Phys. Chem.* **1975**, *79*, 574–577.

(6") 4-Octene, (Z)
     cis-4-Octene
     298 K   28.3(0.1)   118.2(0.4)

The value given is the arithmetic mean of two experiments. Uncertainties are the arithmetic mean of the uncertainties for each experiment which, in turn, are twice the standard deviation from the mean for four degrees of freedom.

Rogers, D. W.; Dejroongruang, K.; Samuel, S. D.; Fang, W.; Zhao, Y. *J. Chem. Thermodynam.* **1992**, *24*, 561–565.

(7)   4-Octene, (E)
     298 K   27.5(0.1)   115.0(0.4)

The value given is the arithmetic mean of two experiments. Uncertainties are the arithmetic mean of the uncertainties for each experiment which, in turn, are twice the standard deviation from the mean for four degrees of freedom.

Rogers, D. W.; Dejroongruang, K.; Samuel, S. D.; Fang, W.; Zhao, Y. *J. Chem. Thermodynam.* **1992**, *24*, 561–565.

(8)   2-Methylhept-1-ene
     298 K   27.5(0.2)   115.1(0.7)

Like the "end effect", (see 2-octene) α methyl substitution stabilizes a simple alkene by about 2.5 kcal mol^{-1}.

Rogers, D. W.; Dejroongruang, K.; Samuel, S. D.; Fang, W.; Zhao, Y. *J. Chem. Thermodynam.* **1992**, *24*, 561–565.

(9)   4-Methylhept-1-ene
        298 K   30.1(0.2)   125.8(0.6)

In simple alkenes, ethyl substitution away from the α position brings about no measurable stabilization.

Rogers, D. W.; Dejroongruang, K.; Samuel, S. D.; Fang, W.; Zhao, Y. *J. Chem. Thermodynam.* **1992**, *24*, 561–565.

(10)   5-Methylhept-1-ene
        298 K   29.8(0.2)   124.6(0.6)

The value given is the arithmetic mean of two experiments. Uncertainties are the arithmetic mean of the uncertainties for each experiment which, in turn, are twice the standard deviation from the mean for four degrees of freedom.

Rogers, D. W.; Dejroongruang, K.; Samuel, S. D.; Fang, W.; Zhao, Y. *J. Chem. Thermodynam.* **1992**, *24*, 561–565.

(11)   6-Methylhept-1-ene
        298 K   30.2(0.5)   126.4(2.0)

The value given is the arithmetic mean of two experiments. Uncertainties are the arithmetic mean of the uncertainties for each experiment which, in turn, are twice the standard deviation from the mean for four degrees of freedom.

Rogers, D. W.; Dejroongruang, K.; Samuel, S. D.; Fang, W.; Zhao, Y. *J. Chem. Thermodynam.* **1992**, *24*, 561–565.

(12)   2,4,4-Trimethylpent-1-ene
        355 K   27.2(0.1)   114.0(0.4)

See also, the commentary on compound (13).

Dolliver, M. A.; Gresham, T. L.; Kistiakowsky, G. B.; Vaughan, W. E. *J. Amer. Chem. Soc.* **1937**, *59*, 831–841.

(12')   2,4,4-Trimethylpent-1-ene
        2,4,4-Trimethyl-1-pentene
        298 K   25.5(0.1)   106.8(0.4)

$\Delta_{hyd}H$ was measured in acetic acid. The discrepancy of 1.7 kcal mol^{-1} between results in acetic acid and gas phase is larger than expected.

Turner, R. B.; Nettleton, D. E.; Perelman, M. *J. Amer. Chem. Soc.* **1958**, *80*, 1430–1433.

(13)   2,4,4-Trimethylpent-2-ene

2,4,4-Trimethyl-2-pentene
355 K   28.4(?)   118.8(?)

The purity of the starting material was questioned by the original authors who subjected the starting material to rigorous analysis and repeated the measurement (see discussion, Dolliver, et al., 1937). Conant and Kistiakowsky (1937) call the result "completely out of line" because the interior double bond, stabilized by a methyl group has a $\Delta_{hyd}H$ that is 1.4 kcal mol^{-1} *larger* than the corresponding compound with a terminal double bond (compound 12). The gas-phase result is supported by G3(MP2) calculations (Rogers, unpublished).

Dolliver, M. A.; Gresham, T. L.; Kistiakowsky, G. B.; Vaughan, W. E. *J. Amer. Chem. Soc.* **1937**, *59*, 831–841.

(13')   2,4,4-Trimethylpent-2-ene
        2,4,4-Trimethyl-2-pentene
        298 K   26.8(0.2)   112.1(0.8)

$\Delta_{hyd}H$ was measured in acetic acid. The discrepancy of 1.6 kcal mol^{-1} between results in acetic acid and the gas-phase result is larger than expected (0.5 - 0.7. kcal mol^{-1}) nevertheless, the anomaly noted in entry (13) was found in these experiments as well. The *exothermic* enthalpy of conversion from the –2– isomer to the –1– isomer found by Kistiakowsky was confirmed in these experiments although the absolute values of $\Delta_{hyd}H$ are different.

Turner, R. B.; Nettleton, D. E.; Perelman, M. *J. Amer. Chem. Soc.* **1958**, *80*, 1430–1433.

(14)   2,5-dimethylhex-3-ene, (Z)
        cis-diisopropylethylene
        298 K   28.7(0.1)   120.0(0.4)

$\Delta_{hyd}H$ was measured in acetic acid solution.

Turner, R. B.; Jarrett, A. D.; Goebel, P.; Mallon, B. J. *J. Amer. Chem. Soc.* **1973**, *95*, 790–792.

(15)   2,5-dimethylhex-3-ene, (E)
        trans-diisopropylethylene
        298 K   26.8(0.1)   112.2(0.4)

$\Delta_{hyd}H$ was measured in acetic acid solution.

Turner, R. B.; Jarrett, A. D.; Goebel, P.; Mallon, B. J. *J. Amer. Chem. Soc.* **1973**, *95*, 790–792.

C_9H_8

(1)   Indene
        Bicyclo[4.3.0]nonatetraene
        355 K   69.9(0.5)   292.5(2.0)

Dolliver, M. A.; Gresham, T. L.; Kistiakowsky, G. B.; Vaughan, W. E. *J. Amer. Chem. Soc.* **1937**, *59*, 831–841.

(2)   Indene
        Bicyclo[4.3.0]nonatetraene
        298 K   23.6(0.3)   98.9(1.2)

Under the mild conditions used, only the nonaromatic double bond was hydrogenated.

Hill, R. K.; Morton, G. H.; Rogers, D. W.; Choi, L. S. *J.Org. Chem.* **1980**, *45*, 5163–5166.

(3)   1-Phenyl-1-propyne
        298 K   62.2(0.5)   260.2(2.0)

Hydrogenation was carried out in *n*-hexane solution. The phenyl group stabilizes the alkyne by 3.4 kcal mol^{-1} relative to methyl stabilization as seen in 2-butyne. The entry above is the arithmetic mean of two separate experiments consisting of 9 hydrogenation runs each.

Davis, H. E.; Allinger, N. L.; Rogers, D. W. *J. Org. Chem.* **1985**, *50*, 3601–3604.

C_9H_{10}

(1)   Indan
        Hydrindene
        Dihydroindene
        355 K   45.8(0.3)   191.6(1.2)

$\Delta_{hyd}H$ of complete hydrogenation of indene minus $\Delta_{hyd}H$ of hydrogenation of indan –69.9(0.5) – (–45.8(0.3)) = –24.1(0.6) should be compared with $\Delta_{hyd}H$ of partial hydrogenation of indene, –23.6(0.3).

Dolliver, M. A.; Gresham, T. L.; Kistiakowsky, G. B.; Vaughan, W. E. *J. Amer. Chem. Soc.* **1937**, *59*, 831–841.

(2)   Allylbenzene
        298 K   30.2(1.0)   126.4(4.0)

The hydrogenation was carried out in cyclohexane. Separation of the benzene ring from the double bond by one carbon atom eliminates phenyl stabilization. See styrene.

Rogers, D. W.; Mc Lafferty, F. J. *Tetrahedron* **1971**, *27*, 3765–3775.

(2')   Allylbenzene
         298 K   30.3(0.2)   126.8(0.8)

The hydrogenation was carried out in glacial acetic acid.

Rogers, D. W.; Mc Lafferty, F. J. *Tetrahedron* **1971**, *27*, 3765–3775.

(3)   Bicyclo[6.1.0]nona-2,4,6-triene
         298 K   120.5(0.1)   504.2(0.4)

The hydrogenation was carried out over 5% Rh/C catalyst, which caused ring opening and yielded cyclononane (see entry immediately following).

Roth, W. R.; Adamczak, O.; Breuckman, R.; Lennartz, H.-W.; Boese, R. *Chem. Ber.* **1991**, *124*, 2499–2521.

(3')   Bicyclo[6.1.0]nona-2,4,6-triene
         298 K   86.0(0.2)   359.8(0.8)

The reaction was carried out over 10% Pd/C catalyst which yielded a simple mixture of cyclononane and bicyclo[6.1.0]nonane. Using the result (3) above, the result of this experiment was calculated for a product of pure bicyclo[6.1.0]nonane.

Roth, W. R.; Adamczak, O.; Breuckman, R.; Lennartz, H.-W.; Boese, R. *Chem. Ber.* **1991**, *124*, 2499–2521.

(4)   2-Phenylpropene
         α-Methylstyrene
         298 K   26.6(0.5)   111.3(2.1)

Hydrogenation was carried out under mild conditions. Only the exocyclic double bond was hydrogenated.

Abboud, J.-L. M.; Jimenez, P.; Roux, V.; Turrion, C.; Lopez-Mardomingo, C.; Podosennin, A.; Rogers, D. W.; Liebman, J. F. *J. Phys. Org. Chem.* **1995**, *8*, 15–25.

(5)   Phenylpropene, (E)
         β-Methylstyrene, (E)
         298 K   25.3(0.3)   105.9(1.3)

Hydrogenation was carried out under mild conditions. Only the exocyclic double bond was hydrogenated.

Abboud, J.-L. M.; Jimenez, P.; Roux, V.; Turrion, C.; Lopez-Mardomingo, C.; Podosennin, A.; Rogers, D. W.; Liebman, J. F. *J. Phys. Org. Chem.* **1995**, *8*, 15–25.

C_9H_{12}

(1) 1,3,5-Trimethylbenzene
     Mesitylene
     355 K   47.6(0.2)   199.2(0.8)

Dolliver, M. A.; Gresham, T. L.; Kistiakowsky, G. B.; Vaughan, W. E *J. Amer. Chem. Soc.* **1937**, *59*, 831–841.

(2) 1,4,7-Cyclononatriene, (E,E,E)
     cis,cis,cis-1,4,7-Cyclononatriene
     298 K   76.9(0.1)   321.7(0.4)

$\Delta_{hyd}H$ was measured in acetic acid solution. The results indicate that homoconjugative stabilization is negligible. See also Turner, R. B.; Meador, W. R. *J. Amer. Chem. Soc.*. **1957**, *79*, 4133–4136 and Turner, R. B.; Mallon, B. J.; Tichy, M.; Doering, W. von E.; Roth. W. R.; Schroeder, G. *J. Amer. Chem. Soc.* **1973**, *95*, 8605–8609.

Roth, W. R.; Bang, W. B.; Goebel, P.; Sass, R. L.; Turner, R. B. *J. Amer. Chem. Soc.* **1964**, *86*, 3178–3179.

(3) Tricyclo[4.2.1.0^{2,5}]non-2(5)-ene
     298 K   44.9(0.2)   187.9(0.8)

No ring cleavage was observed.

Roth, W. R.; Adamczak, O.; Breuckman, R.; Lennartz, H.-W.; Boese, R. *Chem. Ber.* **1991**, *124*, 2499–2521.

(4) 3,3-Dimethylbicyclo[3,2]hepta-1,4-diene
     298 K   69.5(0.4)   290.8(1.6)

Roth, W. R.; Adamczak, O.; Breuckman, R.; Lennartz, H.-W.; Boese, R. *Chem. Ber.* **1991**, *124*, 2499–2521.

(5) 4,4-Dimethylhepta-1,2,5,6-tetraene
     298 K   139.0(0.4)   581.6(1.6)

Roth, W. R.; Adamczak, O.; Breuckman, R.; Lennartz, H.-W.; Boese, R. *Chem. Ber.* **1991**, *124*, 2499–2521.

C_9H_{14}

(1)   Cyclononyne
        298 K   61.9(0.3)   259.1(1.2)

$\Delta_{hyd}H$ was measured in acetic acid solution. Solvent effects are significant for the alkynes and may cause the measured enthalpy changes to be up to 2.0 kcal mol^{-1} less exothermic than the gas-phase values.

Turner, R. B.; Jarrett, A. D.; Goebel, P.; Mallon, B. J. *J. Amer. Chem. Soc.* **1973**, *95*, 790–792.

(2)   1,5-Cyclononadiene, (Z,Z)
        298 K   46.3(0.3)   193.8(1.2)

$\Delta_{hyd}H$ was measured in glacial acetic acid.

Turner, R. B.; Mallon, B. J.; Tichy, M.; Doering, W. von E.; Roth. W. R.; Schroeder, G. *J. Amer. Chem. Soc.* **1973**, *95*, 8605–8609.

(3)   1,5-Cyclononadiene, (Z,E)
        298 K   50.6(0.3)   211.8(1.2)

$\Delta_{hyd}H$ was measured in glacial acetic acid.

Turner, R. B.; Mallon, B. J.; Tichy, M.; Doering, W. von E.; Roth. W. R.; Schroeder, G. *J. Amer. Chem. Soc.* **1973**, *95*, 8605–8609.

C_9H_{16}

(1)   1-Nonyne
        298 K   69.5(0.5)   291.0(2.0)

Rogers, D. W.; Dagdagan, O. A.; Allinger, N. L. *J. Amer. Chem. Soc.* **1979**, *101*, 671–676.

(2)   2-Nonyne
        298 K   65.1(0.5)   272.3(2.0)

The slight trend toward less exothermic enthalpies of hydrogenation for 2-octyne to 4-octyne is due to the larger number of conformers in the alkane product. Thus, $\Delta_f H$ of the alkyne is not *decreasing* along this series, $\Delta_f H$ of the conformational mix of products is *increasing*. Compare the linear octynes and decynes.

Rogers, D. W.; Dagdagan, O. A.; Allinger, N. L. *J. Amer. Chem. Soc.* **1979**, *101*, 671–676.

(3) 3-Nonyne
298 K   64.7(0.3)   270.7(1.3)

Rogers, D. W.; Dagdagan, O. A.; Allinger, N. L. *J. Amer. Chem. Soc.* **1979**, *101*, 671–676.

(4) 4-Nonyne
298 K   64.7(0.4)   270.7(1.7)

Rogers, D. W.; Dagdagan, O. A.; Allinger, N. L. *J. Amer. Chem. Soc.* **1979**, *101*, 671–676.

(5) Cyclononene, (Z)
    cis-Cyclononene
    298 K   23.6(0.1)   98.8(0.4)

$\Delta_{hyd}H$ was measured in acetic acid solution.

Turner, R. B.; Meador, W. R. *J. Amer. Chem. Soc..* **1957**, *79*, 4133–4136. See also Turner, R. B.; Mallon, B. J.; Tichy, M.; Doering, W. von E.; Roth. W. R.; Schroeder, G. *J. Amer. Chem. Soc.* **1973**, *95*, 8605–8609.

(6) Cyclononene, (E)
    trans-Cyclononene
    298 K   26.5(0.1)   110.8(0.4)

$\Delta_{hyd}H$ was measured in acetic acid solution.

Turner, R. B.; Meador, W. R. *J. Amer. Chem. Soc..* **1957**, *79*, 4133–4136. See also Turner, R. B.; Mallon, B. J.; Tichy, M.; Doering, W. von E.; Roth. W. R.; Schroeder, G. *J. Amer. Chem. Soc.* **1973**, *95*, 8605–8609.

(7) 1-Methylcyclooctene, (E)
    298 K   35.2(0.1)   147.3(0.4)

Roth, W. R.; Adamczak, O.; Breuckman, R.; Lennartz, H.-W.; Boese, R. *Chem. Ber.* **1991**, *124*, 2499–2521.

C_9H_{18}

(1) 1-Nonene
    298 K   30.4(0.2)   127.2(0.8)

Hydrogenation was carried out in *n*-hexane. The value differs slightly from the original paper due to a revised standard $\Delta_{hyd}H$, Rogers, D. W., *J. Phys. Chem.* **1979**, *83*, 2430 (one page).

Rogers, D. W.; Skanupong, S. *J. Phys. Chem.* **1974**, *78*, 2569–2572.

$C_{10}H_8$

(1)   Azulene
      Bicyclo[5.3.0]decapentaene
      Cyclopentacycloheptene
      298 K   99.0(0.1)   414.1(0.4)

$\Delta_{hyd}H$ was measured in acetic acid. The authors give a short history of resonance energy calculations for azulene and arrive at their own "crude" (their adjective) estimate of 28 kcal mol^{-1}. Azulene is aromatic because transfer of an electron pair from the seven-membered ring to the five-membered ring causes both rings to be aromatic. It is blue because of this electron transfer, which does not occur in its isomer naphthalene.

Turner, D. W.; Meador, W. R.; Doering, W. von E.; Knox, L. H.; Mayer, J. R.; Wiley, D. W. *J. Amer. Chem. Soc.* **1957**, *79*, 4127–4133.

(2)   Bicyclo[6.2.0]decapentaene
      298 K   135.6(0.8)   567.4(3.2)

$\Delta_{hyd}H$ was measured in cyclohexane. A correction was made of + 0.1 kcal mol^{-1} per double bond in the reactant for the heat of solution difference between reactant and product. Corrections were not made for the difference in heats of vaporization or sublimation in this study. A mixed reaction product was obtained and corrected to the major component using values of $\Delta_f H$ calculated by molecular mechanics. This is reflected in the relatively large experimental uncertainty.

Roth, W. R.; Lennartz, H.-W.; Vogel, E.; Leiendecker, M.; Oda, M.; *Chem. Ber.* **1986**, *119*, 837–843.

$C_{10}H_{10}$

(1)   1,2-Dihydronaphthalene
      302   27.1(0.1)   113.5(0.4)
      355 K (gas phase, estimated)
                           27.6   115.5

Liquid → solution reaction (see note under 1-heptene).

Williams, R. B. *J. Amer. Chem. Soc.* **1942**, *64*, 1395–1404.

(2)   1,4-Dihydronaphthalene
      302   24.1(0.2)   100.8(0.8)

355 K (gas phase, estimated)
  24.6   102.9

Liquid → solution reaction (see note under 1-heptene).

Williams, R. B. *J. Amer. Chem. Soc.* **1942**, *64*, 1395–1404.

(3) 1-Phenyl-1-butyne
     298 K   62.7(0.2)   262.3(0.7)

The hydrogenation was carried out in *n*-hexane solution. The entry above is the arithmetic mean of two separate experiments consisting of 9 hydrogenation runs each.

Davis, H. E.; Allinger, N. L.; Rogers, D. W. *J. Org. Chem.* **1985**, *50*, 3601–3604.

(4) Triquinacene
     tricyclo-[5.2.1.0^{4,10}]deca-2,5,8-triene
     298 K   78.0(0.5)   326.4(2.1)

The hydrogenation was carried out in *n*-hexane solution. These results indicate a small homoaromatic thermochemical stabilization of the target compound. Subsequent work has been mixed on the question of stabilization but negative on the question of homoaromaticity.

Liebman, J. F.; Paquette, L. A.; Peterson, J. R.; Rogers, D. W. *J. Amer. Chem. Soc.* **1986**, *108*, 8267–8268.

(5) 1-Methylene-2-phenylcyclopropane
     298 K   74.14(0.21)   310.2(0.9)

Hydrogenation was carried out in isooctane over 5% Pd/C catalyst. A mixture of 3 phenylbutane isomers differing by only about 2 kcal mol^{-1} in $\Delta_f H$ was corrected to the main product, phenylmethylcyclopropane. The result was corrected for solution effects but not for differences in $\Delta_{vap} H$ between the reactant and product.

Roth, W. R.; Winzer M.; Lennartz, H.-W.; Boese, R. *Chem. Ber.* **1993**, *126*, 2717–2725.

(6) Tricyclo[6.2.0.0^{3,6}]deca-1(8),2,6-triene
     298 K   58.6(0.3)   245.2(1.3)

The result was corrected for the heat of solution difference between reactant and product by 0.1 kcal mol^{-1} per double bond and for the heat of vaporization by the difference between Kovats indices.

Roth, W. R.; Langer, R.; Ebbrecht, T.; Beitat, A.; Lennartz, H.-W. *Chem. Ber.* **1991**, *124*, 2751–2760.

$C_{10}H_{12}$

(1) Cyclododeca-1,2,6,7-tetraene
298 K   123.0(0.2)   514.6(0.8)

Roth, W. R.; Adamczak, O.; Breuckman, R.; Lennartz, H.-W.; Boese, R. *Chem. Ber.* **1991**, *124*, 2499–2521.

(2) 50:50 (E and Z)-Cyclopent-2-enylidinecyclopent-2-ene
298 K   70.5(0.1)   315.9(0.4)

Roth, W. R.; Adamczak, O.; Breuckman, R.; Lennartz, H.-W.; Boese, R. *Chem. Ber.* **1991**, *124*, 2499–2521.

(3) 2-Methyl-3-phenylpropene
    Methallylbenzene
298 K   28.5(0.2)   119.2(0.8)

Hydrogenation was carried out in cyclohexane. The benzene ring is not hydrogenated under mild conditions.

Rogers, D. W.; Mc Lafferty, F. J. *Tetrahedron* **1971**, *27*, 3765–3775.

(4) Tetracyclo[5.2.1.0.^{2,6}0^{3,5}]dec-8-ene
298 K   86.7(0.3)   362.8(1.3)

Roth, W. R.; Klaerner F.-G.; Lennartz, H.-W. *Chem. Ber.* **1980**, *113*, 1818–1829.

(5) Dihydrotriquinacine
298 K   55.0(0.4)   230.1(1.7)

The hydrogenation was carried out in *n*-hexane solution.

Liebman, J. F.; Paquette, L. A.; Peterson, J. R.; Rogers, D. W. *J. Amer. Chem. Soc.* **1986**, *108*, 8267–8268.

(6) 1-Phenyl-2-methylpropene
    β,β-Dimethylstyrene
298 K   25.7(0.4)   107.5(1.7)

Hydrogenation was carried out under mild conditions. Only the exocyclic double bond was hydrogenated.

Abboud, J.-L. M.; Jimenez, P.; Roux, V.; Turrion, C.; Lopez-Mardomingo, C.; Podosennin, A.; Rogers, D. W.; Liebman, J. F. *J. Phys. Org. Chem.* **1995**, *8*, 15–25.

(7) *anti*-(2 + 2)Cyclopentadiene-dimer

298   56.9(0.1)   238.1(0.4)

$\Delta_{hyd}H$ was measured in isooctane solution and corrected for $\Delta_{sol'n}H$ and for $\Delta_{vap}H$ according to the difference in Kovats indices of the reactant and product. An MM correction (see Chapter 3) was made for 5.66% of an impurity presumed to be *endo*-dicyclopentene.

Doering, W. v. E; Roth, W. R.; Breuckman, R.; Figge, L.; Lennartz, H.-W.; Fessner, W.-D.; Prinzbach. H. *Chem. Ber.* **1988**, *121*, 1–9.

(8)   *syn*-(2 + 2)Cyclopentadiene-dimer
      298   54.2   226.8

No error limits are given because only one measurement was made. $\Delta_{hyd}H$ was measured in isooctane solution and corrected for $\Delta_{sol'n}H$ and for $\Delta_{vap}H$ according to the difference in Kovats indices of the reactant and product. Approximately 5% of an impurity was found in the reactant. The $\Delta_{hyd}H$ value is based on hydrogen uptake.

Doering, W. v. E; Roth, W. R.; Breuckman, R.; Figge, L.; Lennartz, H.-W.; Fessner, W.-D.; Prinzbach. H. *Chem. Ber.* **1988**, *121*, 1–9.

(9)   Dispiro[2.2.2.2]deca-4,9-dien
      298 K   106.4(0.2)   445(1.0)

The hydrogenation was carried out in isooctane and corrected for enthalpies of solution, not however for the heat of vaporization. There were three reaction products, propylcyclohexane, isopropylcyclohexane, and methylethylcyclohexane. The error estimate is semiquantitative because the analytical determination of the three reaction products contribute to the error and the enthalpies of formation of the products were calculated by MM2ERW force field calculations.

Roth, W. R.; Unger, C. *Liebigs Ann.* . **1995**, 1361–1366.

$C_{10}H_{14}$

(1)   Tricyclo[4.3.1.0^{1,6}]deca-2,4-diene
      298 K   82.5(0.4)   345.2(1.7)

The methano bridge opens upon hydrogenation to give a mixture of products, with one product dominant (see Tables 7 and 8, original publication.)

Roth, W. R.; Klaerner, F.-G.; Siepert, G.; Lennartz, H.-W. *Chem. Ber.* **1992**, *125*, 217–224.

(2)   Methylenetetramethylcyclopenta-1,3-diene
      298 K   70.8(0.9)   296.2(3.8)

Roth, W. R.; Adamczak, O.; Breuckman, R.; Lennartz, H.-W.; Boese, R. *Chem. Ber.* **1991**, *124*, 2499–2521.

(3) (meso)-cyclopent-2-enylcyclopent-2-ene
    298 K   54.5(0.1)   228.0(0.4)

Roth, W. R.; Adamczak, O.; Breuckman, R.; Lennartz, H.-W.; Boese, R. *Chem. Ber.* **1991**, *124*, 2499–2521.

(4) (racemic)-cyclopent-2-enylcyclopent-2-ene
    298 K   54.6(0.1)   228.4(0.4)

Roth, W. R.; Adamczak, O.; Breuckman, R.; Lennartz, H.-W.; Boese, R. *Chem. Ber.* **1991**, *124*, 2499–2521.

(5) Tricyclo[5.3.0.^{2,5}0]dec-2(5)-ene
    298 K   37.5(0.3)   156.9(1.3)

Roth, W. R.; Adamczak, O.; Breuckman, R.; Lennartz, H.-W.; Boese, R. *Chem. Ber.* **1991**, *124*, 2499–2521.

(6) Tetrahydrotriquinacene
    298 K   27.5(0.3)   115.1(1.3)

The hydrogenation was carried out in *n*-hexane solution.

Liebman, J. F.; Paquette, L. A.; Peterson, J. R.; Rogers, D. W. *J. Amer. Chem. Soc.* **1986**, *108*, 8267–8268.

$C_{10}H_{16}$

(1) 2-Methyl-5-isopropylcyclohexa-1,3-diene
    α-Phellandrene
    355 K   53.4(0.3)   223.5(1.3)

The reactant was of doubtful purity.

Dolliver, M. A.; Gresham, T. L.; Kistiakowsky, G. B.; Vaughan, W. E. *J. Amer. Chem. Soc.* **1937**, *59*, 831–841.

(2) 1-Methyl-4-isopropylcyclohexa-1,3-diene
    α-Terpinene
    355 K   50.7(0.3)   212.1(1.3)

The reactant was of doubtful purity.

Dolliver, M. A.; Gresham, T. L.; Kistiakowsky, G. B.; Vaughan, W. E. *J. Amer. Chem. Soc.* **1937**, *59*, 831–841.

(3) 1-Methyl-4-isopropenylcyclohex-1-ene
    Limonene
    355 K   54.1(0.3)   226.4(1.3)

The reactant was of doubtful purity.

Dolliver, M. A.; Gresham, T. L.; Kistiakowsky, G. B.; Vaughan, W. E. *J. Amer. Chem. Soc.* **1937**, *59*, 831–841.

(4) 3-decen-1-yne, (E)
    trans-dec-3-en-1-yne
    298 K   96.0(0.2)   401.7(0.8)

$\Delta_{hyd}H$ was measured in glacial acetic acid and corrected by a separate heat of solution measurement experiment. Value given for the liquid → liquid reaction.

Skinner, H. A.; Snelson A. *Trans. Faraday Soc.* **1959**, *55*, 404–407.

(5) 3-decen-1-yne, (Z)
    cis-dec-3-en-1-yne
    298 K   95.6(0.5)   400.0(2.1)

$\Delta_{hyd}H$ was measured in glacial acetic acid and corrected by a separate heat of solution measurement experiment. Value given for the liquid → liquid reaction.

Skinner, H. A.; Snelson A. *Trans. Faraday Soc.* **1959**, *55*, 404–407.

(6) Cyclodecyne
    298 K   56.5(0.2)   236.2(0.8)

$\Delta_{hyd}H$ was measured in acetic acid solution. Solvent effects are significant for the alkynes and may cause the measured enthalpy changes to be up to 2.0 kcal mol^{-1} less exothermic than the gas-phase values.

Turner, R. B.; Jarrett, A. D.; Goebel, P.; Mallon, B. J. *J. Amer. Chem. Soc.* **1973**, *95*, 790–792.

See also Sicher, J.; Svoboda, M.; Zavada, J.; Turner, R. B.; Goebel, P. *Tetrahedron*, **1966**, *22*, 659–671.

(7) Cyclodeca-1,6-diene, (Z,Z)
    298 K   43.7   182.8

$\Delta_{hyd}H$ was measured in glacial acetic acid. The sample purity was questionable. No experimental uncertainty was given.

Turner, R. B.; Mallon, B. J.; Tichy, M.; Doering, W. von E.; Roth. W. R.; Schroeder, G. *J. Amer. Chem. Soc.* **1973**, *95*, 8605–8609.

(8) Cyclodeca-1,6-diene, (E,E)
     298 K   47.6   199.3

$\Delta_{hyd}H$ was measured in glacial acetic acid. The sample purity was questionable.

Turner, R. B.; Mallon, B. J.; Tichy, M.; Doering, W. von E.; Roth. W. R.; Schroeder, G. *J. Amer. Chem. Soc.* 8605–8609.

(7) Tricyclo[4.3.1.0^{1,6}]dec-2-ene
     298 K   60.0(0.1)   251.0(0.4)

The methano bridge opens upon hydrogenation to give a mixture of products, with one product dominant (see Tables 7 and 8, original publication.)

Roth, W. R.; Klaerner, F.-G.; Siepert, G.; Lennartz, H.-W. *Chem. Ber.* **1992**, *125*, 217–224.

(8) Tricyclo[4.3.1.0^{1,6}]dec-3-ene
     298 K   59.1(0.1)   247.3(0.4)

The methano bridge opens upon hydrogenation to give a mixture of products, with one product dominant (see Tables 7 and 8, original publication.)

Roth, W. R.; Klaerner, F.-G.; Siepert, G.; Lennartz, H.-W. *Chem. Ber.* **1992**, *125*, 217–224.

(9) Pentamethylcyclopenta-1,3-diene
     298 K   45.0(0.2)   188.3(0.8)

Roth, W. R.; Adamczak, O.; Breuckman, R.; Lennartz, H.-W.; Boese, R. *Chem. Ber.* **1991**, *124*, 2499–2521.

$C_{10}H_{18}$

(1) 1-Decyne
     298 K   69.6(0.5)   291.4(2.0)

Rogers, D. W.; Dagdagan, O. A.; Allinger, N. L. *J. Amer. Chem. Soc.* **1979**, *101*, 671–676.

(2) 2-Decyne
     298 K   65.3(0.5)   273.1(2.0)

The slight trend toward less exothermic enthalpies of hydrogenation for 2-octyne to 4-octyne is due to the larger number of conformers in the alkane product. Thus, $\Delta_f H$ of the alkyne is not *decreasing* along this series, $\Delta_f H$ of the conformational mix of products is *increasing*. Compare the linear octynes and nonynes.

Rogers, D. W.; Dagdagan, O. A.; Allinger, N. L. *J. Amer. Chem. Soc.* **1979**, *101*, 671–676.

(3) 3-Decyne
     298 K   64.9(0.5)   271.4(2.0)

Rogers, D. W.; Dagdagan, O. A.; Allinger, N. L. *J. Amer. Chem. Soc.* **1979**, *101*, 671–676.

(4) 4-Decyne
     298 K   64.4(0.4)   269.4(1.7)

Rogers, D. W.; Dagdagan, O. A.; Allinger, N. L. *J. Amer. Chem. Soc.* **1979**, *101*, 671–676.

(5) 5-Decyne
     298 K   64.1(0.5)   268.2(2.0)

Rogers, D. W.; Dagdagan, O. A.; Allinger, N. L. *J. Amer. Chem. Soc.* **1979**, *101*, 671–676.

(6) Cyclodecene, (Z)
     cis-Cyclodecene
     298 K   20.7(0.1)   86.5(0.4)

$\Delta_{hyd} H$ was measured in acetic acid solution. See also Turner, R. B.; Mallon, B. J.; Tichy, M.; Doering, W. von E.; Roth. W. R.; Schroeder, G. *J. Amer. Chem. Soc.* **1973**, *95*, 8605–8609 and Sicher, J.; Svoboda, M.; Zavada, J.; Turner, R. B.; Goebel, P. *Tetrahedron*, **1966**, *22*, 659–671.

Turner, R. B.; Meador, W. R. *J. Amer. Chem. Soc.*. **1957**, *79*, 4133–4136.

(7) Cyclodecene, (E)
     trans-Cyclodecene
     298 K   24.0(0.1)   100.5(0.4)

$\Delta_{hyd} H$ was measured in acetic acid solution. See also Turner, R. B.; Mallon, B. J.; Tichy, M.; Doering, W. von E.; Roth. W. R.; Schroeder, G. *J. Amer. Chem. Soc.* **1973**,

95, 8605 – 8609 and Sicher, J.; Svoboda, M.; Zavada, J.; Turner, R. B.; Goebel, P. *Tetrahedron*, **1966**, *22*, 659–671.

Turner, R. B.; Meador, W. R. *J. Amer. Chem. Soc.*. **1957**, *79*, 4133 – 4136.

(8)   2,2,5,5-Tetramethylhex-3-yne
        298 K   66.7(0.1)   279.1(0.4)

Roth, W. R.; Hopf, H.; Horn, C. *Chem. Ber.* **1994**, *127*, 1781–1795. $\Delta_{hyd}H$ was measured in isooctane. The datum is not corrected for either $\Delta_{vap}H$ or $\Delta_{sol'n}H$.

$C_{10}H_{20}$

(1)   1-Decene
        299.1   29.9(0.3)   125.1(1.3)

The hydrogenation was carried out in an oscillatory calorimeter using glacial acetic acid as the calorimeter fluid. Results are corrected for $\Delta_{sol'n}H$ of the product in acetic acid.

Bretschneider, E.; Rogers, D. W. *Mikrochem. Acta [Wien]* **1970**, 482–490.

(1')   1-Decene
        298 K   30.4(0.2)   127.1(1.3)

The hydrogenation was carried out in *n*-hexane. The value differs slightly from the original paper due to a revised standard $\Delta_{hyd}H$, Rogers, D. W., *J. Phys. Chem.* **1979**, *83*, 2430 (one page).

Rogers, D. W.; Skanupong, S. *J. Phys. Chem.* **1974**, *78*, 2569–2572.

(2)   5-Decene
        298 K   29.3(0.4)   122.6(1.7)

The hydrogenation was carried out in *n*-hexane. Value differs slightly from the original paper due to a revised standard $\Delta_{hyd}H$, Rogers, D. W., *J. Phys. Chem.* **1979**, *83*, 2430 (one page).

Rogers, D. W.; Siddiqui, N. A. *J. Phys. Chem.* **1975**, *79*, 574 – 577.

(3)   2,2,5,5-Tetramethylhex-3-ene, (Z)
        cis-di(*t*-butyl)-ethylene
        298 K   36.2(0.2)   151.6(0.8)

$\Delta_{hyd}H$ was measured in acetic acid.

Turner, R. B.; Nettleton, D. E.; Perelman, M. *J. Amer. Chem. Soc.* **1958**, *80*, 1430–1433.

(3')   2,2,5,5-Tetramethylhex-3-ene, (Z)
       di(t-butyl)-ethylene
       298 K   37.7(0.1)   157.7(0.4)

von E. Doering, W.; Roth, W. R.; Bauer, F.; Boenke, M.; Breuckmann, R.; Ebbrecht, M; Herbold, M.; Schmidt, R.; Herbold, M.; Schmidt, R.; Herbold, M.; Schmidt, R.; Lennartz, H.-W.; Lenoir, D.; Boese, R. *Chem. Ber.* **1989**, *122*, 1263–1275.

(4)   2,2,5,5-Tetramethylhex-3-ene, (E)
       trans-di(t-butyl)-ethylene
       298 K   26.9(0.1)   112.4(0.4)

$\Delta_{hyd}H$ was measured in acetic acid.

Turner, R. B.; Nettleton, D. E.; Perelman, M. *J. Amer. Chem. Soc.* **1958**, *80*, 1430–1433.

(4')   2,2,5,5-Tetramethylhex-3-ene, (E)
       di(t-butyl)-ethylene
       298 K   28.1(0.2)   117.6(0.8)

von E. Doering, W.; Roth, W. R.; Bauer, F.; Boenke, M.; Breuckmann, R.; Ebbrecht, M; Herbold, M.; Schmidt, R.; Lennartz, H.-W.; Lenoir, D.; Boese, R. *Chem. Ber.* **1989**, *122*, 1263–1275.

(5)   2,2,4,4-tetramethyl-3-methylenepentane
       1,1-di(t-butyl)ethylene
       298 K   28.0(0.1)   117.2(0.4)

$\Delta_{hyd}H$ was measured in acetic acid solution. Solvent effects are significant for the alkenes and may cause the measured enthalpy changes to be ~ 0.7 kcal mol^{-1} less exothermic than the gas-phase values. The error is larger for polyunsaturated compounds. Compare 2-methylbut-2-ene, Kistiakowsky, G. B.; Ruhoff, J. R.; Smith, H. A.; Vaughan, W. E. *J. Amer. Chem. Soc.* **1936**, *58*, 137–145.

Turner, R. B.; Jarrett, A. D.; Goebel, P.; Mallon, B. J. *J. Amer. Chem. Soc.* **1973**, *95*, 790–792.

$C_{11}H_{10}$

(1)   Tricyclo[4.4.1^{9,10}.0^{1,6}]undeca-1,2,4,7-tetraene
       298 K   104.0(0.4)   435.1(1.7)

Hydrogenation yields a complex mixture of products. See Tables 7 and 8 in the original publication.

Roth, W. R.; Klaerner, F.-G.; Siepert, G.; Lennartz, H.-W. *Chem. Ber.* **1992**, *125*, 217–224.

(2)   Bicyclo[4.4.1]undeca-1,3,5,7,9-pentaene
       298 K   120.7(0.1)   505.0(0.4)

Hydrogenation yields a complex mixture of products. See Tables 7 and 8 in the original publication.

Roth, W. R.; Klaerner, F.-G.; Siepert, G.; Lennartz, H.-W. *Chem. Ber.* **1992**, *125*, 217–224.

$C_{11}H_{12}$

(1)   Bicyclo[4.4.1]undeca-1,3,5,8-tetraene
       298 K   107.7(0.1)   450.6(0.4)

Hydrogenation yields a complex mixture of products. See Tables 7 and 8 in the original publication.

Roth, W. R.; Klaerner, F.-G.; Siepert, G.; Lennartz, H.-W. *Chem. Ber.* **1992**, *125*, 217–224.

(2)   Tricyclo[4.4.1^{9,10}.0^{1,6}]undeca-1,2,4-triene
       298 K   81.7(0.2)   341.8(0.8)

Hydrogenation yields a complex mixture of products. See Tables 7 and 8 in the original publication.

Roth, W. R.; Klaerner, F.-G.; Siepert, G.; Lennartz, H.-W. *Chem. Ber.* **1992**, *125*, 217–224.

$C_{11}H_{14}$

(1)   Bicyclo[4.4.1]undeca-1,3,5-triene
       298 K   72.1(0.1)   301.7(0.4)

Hydrogenation yields a complex mixture of products. See Tables 7 and 8 in the original publication.

Roth, W. R.; Klaerner, F.-G.; Siepert, G.; Lennartz, H.-W. *Chem. Ber.* **1992**, *125*, 217–224.

(2)   [5]Metacyclophane
       m-Pentanobenzene

298 K	52.6(1)	220.(4)

The datum is the arithmetic mean of three separate experiments of 8, 9, and 8 hydrogenations, one using 1-hexene and the other using styrene as thermochemical standards. The uncertainties are 95% confidence limits. The reaction is partial, due to kinetic lag. In this reaction the normally facile hydrogenation of a monoalkene and the reluctance of a benzene ring to react under mild conditions are reversed by strain. Two moles of hydrogen are taken up by the torsionally strained benzene ring to produce the hindered bridgehead double bond in the first reaction product, bicyclo[5.3.1]undec-1(10)-ene. With time the expected alkane, bicyclo[5.3.1]undecane, is produced but this second step is not seen in the $\Delta_{hyd}H$ shown here.

van Eis, M. J.; Wijsman, G. W.; de Wolf, W. H.; Bickelhaupt, F.; Rogers, D. W.; Kooijman, H.; Spek, A. L. *Chem. Eur. J.* **2000**, *6*, 1537–1546.

$C_{11}H_{16}$

(1) Tricyclo[4.3.2.0^{1,6}]undeca-2,4-diene
298 K   49.3(0.2)   206.3(0.8)

Roth, W. R.; Klaerner, F.-G.; Siepert, G.; Lennartz, H.-W. *Chem. Ber.* **1992**, *125*, 217–224.

$C_{11}H_{18}$

(1) Cycloundecyne
298 K   57.2(0.2)   239.3(0.8)

$\Delta_{hyd}H$ was measured in acetic acid solution. Solvent effects are significant for the alkynes and may cause the measured enthalpy changes up to 2.0 kcal mol^{-1} less exothermic than the gas-phase values.

Turner, R. B.; Jarrett, A. D.; Goebel, P.; Mallon, B. J. *J. Amer. Chem. Soc.* **1973**, *95*, 790–792.

(2) Tricyclo[4.3.2.0^{1,6}]undec-3-ene
298 K   25.1(0.1)   105.0(0.4)

Roth, W. R.; Klaerner, F.-G.; Siepert, G.; Lennartz, H.-W. *Chem. Ber.* **1992**, *125*, 217–224.

$C_{11}H_{22}$

(1)   1-Undecene
        298 K   30.3(0.4)   126.6(1.7)

The hydrogenation was carried out in *n*-hexane. The value differs slightly from the original paper due to a revised standard $\Delta_{hyd}H$ Rogers, D. W., *J. Phys. Chem.* **1979**, *83*, 2430 (one page).

Rogers, D. W.; Skanupong, S. *J. Phys. Chem.* **1974**, *78*, 2569 – 2572.

$C_{12}H_{12}$

(1)   Tricyclo[4.4.2.$1^{,6}0^{1,6}$]dodeca-2,4,7,9-tetraene
        1,6-Ethanonaphthalene
        298 K   106.1(0.1)   433.9(0.4)

Roth, W. R.; Klaerner, F.-G.; Siepert, G.; Lennartz, H.-W. *Chem. Ber.* **1992**, *125*, 217–224.

$C_{12}H_{14}$

(1)   *exo,exo*-Tetracyclo[6.2.1.1$_{2,5}^{3,6}$.1$_{7,10}^{2,7}$0]dodeca-4,9-diene
        syn-Tetracyclo[4.4.0.1.$^{\ }$1$^{\ }$]dodeca-3,8-diene
        syn-Dinorbornene
        298 K   61.0(0.3)   255.4(1.3)

$\Delta_{hyd}H$ was measured in acetic acid solution. Solvent effects are significant for the alkenes and may cause the measured enthalpy changes to be ~ 0.7 kcal mol^{-1} less exothermic than the gas-phase values. Interest in this and the following dimethanonaphthalene derivatives arises from their enthalpies of partial hydrogenation which is, sequentially, -34.1 and –26.9 kcal mol^{-1}. The unusually small value of $\Delta_{hyd}H$ for the second step is indicative of repulsion between the ethano bridges in the tetracyclo[4.4.1.1.0]dodecane produced.

Turner, R. B.; Jarrett, A. D.; Goebel, P.; Mallon, B. J. *J. Amer. Chem. Soc.* **1973**, *95*, 790–792.

(2)   *exo,endo*-Tetracyclo[6.2.1.1$_{2,5}^{3,6}$.1$_{7,10}^{2,7}$0]dodeca-4,9-diene
        anti-Tetracyclo[4.4.0.1.$^{\ }$1$^{\ }$]dodeca-3,8-diene
        anti-Dinorbornene
        298 K   69.2(0.3)   289.4(1.3)

$\Delta_{hyd}H$ was measured in acetic acid solution. Solvent effects are significant for the alkenes and may cause the measured enthalpy changes to be ~ 0.7 kcal mol^{-1} less exothermic than the gas-phase values. Compare norbornene, Turner, R. B.; Meador, W. R.; Winkler, R. E. *J. Amer. Chem. Soc.* **1957**, *79*, 4116–4121 and Rogers, D. W.; Choi, L.

S.; Girellini, R. S.; Holmes, T. J.; Allinger, N. L., *J. Phys. Chem.* **1980**, *84*, 1810 – 1814. The stepwise hydrogenation of this diene is more nearly normal in its $\Delta_{hyd}H$, as seen by comparison with norbornene. See however dihydronorbornenes (2) and (3) immediately below.

Turner, R. B.; Jarrett, A. D.; Goebel, P.; Mallon, B. J. *J. Amer. Chem. Soc.* **1973**, *95*, 790–792.

(4)   Tricyclo[4.4.2.^{1,6}0^{1,6}]dodeca-2,4,8-triene
        1,6-Ethanodihydronaphthalene
        298 K   81.8(0.1)   342.3(0.4)

Roth, W. R.; Klaerner, F.-G.; Siepert, G.; Lennartz, H.-W. *Chem. Ber.* **1992**, *125*, 217–224.

(5)   1-Phenyl-1-hexyne
        298   62.6(0.3)   261.7(1.3)

The hydrogenation was carried out in *n*-hexane solution. The triple bond is stabilized by a phenyl group on one end and an alkyl group on the other. The entry above is the arithmetic mean of two separate experiments consisting of 9 hydrogenation runs each.

Davis, H. E.; Allinger, N. L.; Rogers, D. W. *J. Org. Chem.* **1985**, *50*, 3601–3604.

(6)   1-Phenylcyclohexene
        298 K   25.8(0.2)   107.9(0.8)

Hydrogenation was carried out in *n*-hexane using styrene ($\Delta_{hyd}H$ = -28.20 kcal mol^{-1}) as a thermochemical standard. The tabulated value is the arithmetic mean of two separate experiments, each consisting of 9 runs of sample *vs.* standard. The uncertainty is estimated to express 95% confidence limits.

$C_{12}H_{16}$

(1)   *syn*-Tetracyclo[6.2.1.1.^{3,6}$_{2,5}$0^{2,7}$_{7,10}$]dodeca-4-ene
        syn-Tetracyclo[4.4.0.1.$^{}$1^{7,10}]dodeca-3-ene
        syn-Dihydrodinorbornene
        298 K   26.9(0.1)   112.5(0.4)

$\Delta_{hyd}H$ was measured in acetic acid solution. Solvent effects are significant for the alkenes and may cause he measured enthalpy changes to be ~ 0.7 kcal mol^{-1} less exothermic than the gas-phase values.

Turner, R. B.; Jarrett, A. D.; Goebel, P.; Mallon, B. J. *J. Amer. Chem. Soc.* **1973**, *95*, 790–792.

(2)   anti-exo-Tetracyclo[6.2.1.1.^{3,6}0^{2,7}]dodeca-4-ene

anti-exo-Tetracyclo[4.4.0.1.^{2,5}1^{7,10}]dodeca-3-ene
anti-exo-Dihydrodinorbornene
298 K   35.4(0.1)   147.9(0.4)

$\Delta_{hyd}H$ was measured in acetic acid solution. Solvent effects are significant for the alkenes and may cause the measured enthalpy changes to be ~ 0.7 kcal mol^{-1} less exothermic than the gas-phase values. This compound has juxtaposition between the methano and ethnao bridges.

Turner, R. B.; Jarrett, A. D.; Goebel, P.; Mallon, B. J. *J. Amer. Chem. Soc.* **1973**, **95**, 790–792.

(3)   anti-endo-Tetracyclo[6.2.1.1.^{3,6}0^{2,7}]dodeca-4-ene

anti-endo-Tetracyclo[4.4.0.1.^{2,5}1^{7,10}]dodeca-3-ene
anti-endo-Dihydrodinorbornene
298 K   31.9(0.2)   133.6(0.8)

$\Delta_{hyd}H$ was measured in acetic acid solution. Solvent effects are significant for the alkenes and may cause the measured enthalpy changes to be ~ 0.7 kcal mol^{-1} less exothermic than the gas-phase values. This compound has juxtaposition between the methano and etheno bridges.

Turner, R. B.; Jarrett, A. D.; Goebel, P.; Mallon, B. J. *J. Amer. Chem. Soc.* **1973**, **95**, 790–792.

(4)   Cyclohex-2-enylidinecyclohex-2-ene, (E)
trans-Cyclohex-2-enylidinecyclohex-2-ene
(E) 3,3'-Bis-(1-cyclohexenylidene)
298 K   73.4(0.2)   307.1(0.8)

Calculated for 100% reduction to dicyclohexane.

Roth, W. R.; Adamczak, O.; Breuckman, R.; Lennartz, H.-W.; Boese, R. *Chem. Ber.* **1991**, *124*, 2499–2521.

(5)   Cyclohex-2-enylidinecyclohex-2-ene, (Z)
cis-Cyclohex-2-enylidinecyclohex-2-ene
(Z) 3,3'-Bis-(1-cyclohexenylidene)
298 K   73.7(0.2)   308.4(0.8)

Calculated for 100% reduction to dicyclohexane.

Roth, W. R.; Adamczak, O.; Breuckman, R.; Lennartz, H.-W.; Boese, R. *Chem. Ber.* **1991**, *124*, 2499–2521.

(6)   syn-Tricyclo[4.2.2.2.^{2,5}]dodeca-3,7-diene
        syn-Tetrahydro[2.2]paracyclophane
        298 K   55.5(0.2)   232.2(0.8)

Roth, W. R.; Adamczak, O.; Breuckman, R.; Lennartz, H.-W.; Boese, R. *Chem. Ber.* **1991**, *124*, 2499–2521.

(7)   anti-Tricyclo[4.2.2.2.^{2,5}]dodeca-3,7-diene
        anti-Tetrahydro[2.2]paracyclophane
        298 K   52.9   221.3

No experimental uncertainty was given because only one measurement was made.

Roth, W. R.; Adamczak, O.; Breuckman, R.; Lennartz, H.-W.; Boese, R. *Chem. Ber.* **1991**, *124*, 2499–2521.

(8)   Tricyclo[4.4.2.0^{1,6}]dodeca-2,4-diene
        298 K   53.3(0.1)   223.0(0.4)

Roth, W. R.; Klaerner, F.-G.; Siepert, G.; Lennartz, H.-W. *Chem. Ber.* **1992**, *125*, 217–224.

(9)   [6]Metacyclophane
        m-Hexanobenzene
        298 K   33.8(1)   141(4)

The datum is the arithmetic mean of two separate experiments of 9 hydrogenations each, one using 1-hexene and the other using styrene as thermochemical standards. The uncertainties are 95% confidence limits. The rection is partial, due to kinetic lag. In this reaction the normally facile hydrogenation of a monoalkene and the reluctance of a benzene ring to react under mild conditions are reversed by strain. Two moles of hydrogen are taken up by the torsionally strained benzene ring to produce two kinetically hindered bridgehead monoalkenes in the first reaction product, bicyclo[6.3.1]dodeca-1(10)-ene and bicyclo[6.3.1]dodeca-1(12)-ene. With time, the expected bicyclo[6.3.1]dodecane, is produced but this second step is not seen in the $\Delta_{hyd}H$ shown here.

van Eis, M. J.; Wijsman, G. W.; de Wolf, W. H.; Bickelhaupt, F. Rogers, D. W.; Kooijman, H.; Spek, A. L. *Chem. Eur. J.* **2000**, *6*, 1537–1546.

$C_{12}H_{18}$

(1)   Dodeca-3,9-diyne
        298 K   131.2(0.5)   548.9(2.0)

$\Delta_{hyd}H$ was measured in glacial acetic acid and corrected for heat of solution. The value given is for a liquid → liquid reaction.

Flitcroft, T. L,; Skinner, H. A.; Whiting, M. C. *Trans. Faraday Soc.* **1957**, *53*, 784–790.

(2) Dodeca-5,7-diyne
298 K   127.3(0.7)   532.6(3.0)

$\Delta_{hyd}H$ was measured in glacial acetic acid and corrected for heat of solution. The value given is for a liquid → liquid reaction.

Flitcroft, T. L.; Skinner, H. A.; Whiting, M. C. *Trans. Faraday Soc.* **1957**, *53*, 784–790.

(3) 1,2,3,4,5,6-Hexamethylbicyclo[2.2.0]hexa-2,5-diene
Hexamethyl(dewarbenzene)
298 K   31.4(0.7)   131.4(3.0)

Hydrogenation was carried out in cyclohexane. The bicyclic structure is retained and only one double bond is hydrogenated, yielding 1,2,3,4,5*endo*,6*endo*-hexamethylbicyclo[2.2.0] hex-2-ene.

Rogers, D. W.; Mc Lafferty, F. J. *Tetrahedron* **1971**, *27*, 3765–3775.

(4) Cyclohex-2-enylcyclohex-2-ene
3,3'-Dicyclohexene
298 K   57.4(0.2)   240.2(0.8)

Roth, W. R.; Adamczak, O.; Breuckman, R.; Lennartz, H.-W.; Boese, R. *Chem. Ber.* **1991**, *124*, 2499–2521.

(5) Tricyclo[4.2.2.2^{2,5}]dodec-3-ene
Octahydro[2.2]paracyclophane
298 K   26.7(0.2)   111.7(0.8)

Roth, W. R.; Adamczak, O.; Breuckman, R.; Lennartz, H.-W.; Boese, R. *Chem. Ber.* **1991**, *124*, 2499–2521.

$C_{12}H_{20}$

(1) Cyclododecyne
298 K   61.7(0.4)   258.4(1.7)

$\Delta_{hyd}H$ was measured in acetic acid solution. Solvent effects are significant, and may cause the measured enthalpy change to be as much as 2.0 kcal mol^{-1} less exothermic than the gas-phase values.

Turner, R. B.; Jarrett, A. D.; Goebel, P.; Mallon, B. J. *J. Amer. Chem. Soc.* **1973**, *95*, 790–792.

See also Sicher, J.; Svoboda, M.; Mallon, B. J.; Turner, R. B. *J. Chem. Soc. (B)*, **1968**, 441–447.

$C_{12}H_{22}$

(1)   Cyclododecene, (E)
      trans-Cyclododecene
      298 K   26.8   112.1

$\Delta_{hyd}H$ was measured in acetic acid solution. An error estimate was not given.

Sicher, J.; Svoboda, M.; Mallon, B. J.; Turner, R. B. *J. Chem. Soc.* (B), 1968, 441–447.

(2)   Cyclododecene, (Z)
      cis-Cyclododecene
      298 K   26.3   110.0

$\Delta_{hyd}H$ was measured in acetic acid solution. The error estimate was not given.

Sicher, J.; Svoboda, M.; Mallon, B. J.; Turner, R. B. *J. Chem. Soc. (B)*, **1968**, 441–447.

(3)   2,2,7,7,Tetramethyloct-4-yne
      298 K   62.4(0.2)   261.1(0.8)

Roth, W. R.; Hopf, H.; Horn, C. *Chem. Ber.* **1994**, *127*, 1781–1795. $\Delta_{hyd}H$ was measured in isooctane. The datum is not corrected for either $\Delta_{vap}H$ or $\Delta_{sol'n}H$.

(4)   2,3-Di(*t-butyl*)-buta-1,3-diene
      298 K   56.8(0.2)   237.9(0.8)

The hydrogenation was carried out in isooctane and corrected for the double bond increment in heat of solution and for estimated differences in $\Delta_{vap}H$ (see original publication, Table IV).

Roth, W. R.; Lennartz, H.-W.; von E. Doering; Dolbier Jr., W. R.; Schmidhauser, J. C. *J. Amer. Chem. Soc.* **1988**, *110*, 1883–1889.

$C_{12}H_{24}$

(1)   1-Dodecene
      299.1   30.1(0.5)   125.7(2.0)

The hydrogenation was carried out in an oscillatory calorimeter using glacial acetic acid as the calorimeter fluid. The results are corrected for $\Delta_{sol'n}H$ of the product in acetic acid. $\Delta_{sol'n}H$ is quite significant for the higher alkanes, being 2.1(0.1) kcal mol^{-1} in this case.

Bretschneider, E.; Rogers, D. W. *Mikrochem. Acta [Wien]* **1970**, 482–490.

(1')   1-Dodecene
        298 K   30.7(0.4)   128.4(1.7)

Hydrogenation was carried out in *n*-hexane. The value differs slightly from the original paper due to a revised standard $\Delta_{hyd}H$, Rogers, D. W., *J. Phys. Chem.* **1979**, *83*, 2430 (one page).

Rogers, D. W.; Skanupong, S. *J. Phys. Chem.* **1974**, *78*, 2569–2572.

(2)   2,2,7,7-tetramethyloct-4-ene, (Z)
        cis-1,2-dineopentylethylene
        298 K   26.9(0.1)   112.5(0.4)

$\Delta_{hyd}H$ was measured in acetic acid solution. Solvent effects are significant and may cause the measured enthalpy changes to be ~ 0.7 kcal mol^{-1} less exothermic than the gas-phase values but the relative difference between E and Z isomers should be the same because the solvent effects cancel. Compare the paradigmatic case of E and Z 2-butene.

Turner, R. B.; Jarrett, A. D.; Goebel, P.; Mallon, B. J. *J. Amer. Chem. Soc.* **1973**, *95*, 790–792.

(3)   2,2,7,7-tetramethyloct-4-ene, (E)
        trans-1,2-dineopentylethylene
        298 K   26.0(0.15)   109.0(0.6)

$\Delta_{hyd}H$ was measured in acetic acid solution. Solvent effects are significant and may cause the measured enthalpy changes to be ~ 0.7 kcal mol^{-1} less exothermic than the gas-phase values.

Turner, R. B.; Jarrett, A. D.; Goebel, P.; Mallon, B. J. *J. Amer. Chem. Soc.* **1973**, *95*, 790–792.

(4)   3,3,8,8-Tetramethylcyclooctyne
        298 K   73.7(0.1)   308.4(0.4)

Roth, W. R.; Hopf, H.; Horn, C. *Chem. Ber.* **1994**, *127*, 1781–1795. $\Delta_{hyd}H$ was measured in isooctane. The datum is not corrected for either $\Delta_{vap}H$ or $\Delta_{sol'n}H$.

(5)   4,4,7,7-Tetramethylcyclooctyne

298 K   72.7(0.1)   304.2(0.4)

Roth, W. R.; Hopf, H.; Horn, C. *Chem. Ber.* **1994**, *127*, 1781–1795. $\Delta_{hyd}H$ was measured in isooctane. The datum is not corrected for either $\Delta_{vap}H$ or $\Delta_{sol'n}H$.

(6)   2,2,3,4,5,5-Hexamethylhex-3-ene, (Z)
       di(*t*-butyl)-*cis*-dimethylethylene
       298 K   43.7(0.2)   182.8(0.8)

The enthalpy of hydrogenation was corrected for small amounts of geometric rearrangement as detected in the product.

von E. Doering, W.; Roth, W. R.; Bauer, F.; Boenke, M.; Breuckmann, R.; Ebbrecht, M; Herbold, M.; Schmidt, R.; Lennartz, H.-W.; Lenoir, D.; Boese, R. *Chem. Ber.* **1989**, *122*, 1263–1275.

(7)   2,2,3,4,5,5-Hexamethylhex-3-ene, (E)
       di(*t*-butyl)-*trans*-dimethylethylene
       298 K   37.4(0.2)   156.5(0.8)

The enthalpy of hydrogenation was corrected for small amounts of geometric rearrangement as detected in the product.

von E. Doering, W.; Roth, W. R.; Bauer, F.; Boenke, M.; Breuckmann, R.; Ebbrecht, M; Herbold, M.; Schmidt, R.; Lennartz, H.-W.; Lenoir, D.; Boese, R. *Chem. Ber.* **1989**, *122*, 1263–1275.

$C_{13}H_{16}$

(1)   1-Phenylcycloheptene
       298 K   23.8(0.2)   99.6(0.8)

Hydrogenation was carried out in *n*-hexane using styrene ($\Delta_{hyd}H = -28.20$ kcal mol^{-1}) as a thermochemical standard. The tabulated value is the arithmetic mean of two separate experiments, each consisting of 9 runs of sample *vs.* standard. The uncertainty is estimated to express 95% confidence limits.

$C_{13}H_{21}$

(1)   *trans*-2-Methylene-7-isopropylbicyclo[4.3.0]nonane
       298 K   27.3(0.4)   114.2(1.7)

Hydrogenation was carried out in *n*-hexane.

Rogers, D. W., Unpublished.

(2)   trans-2-Methyl-7-isopropylbicyclo[4.3.0]non-2-ene
        298 K   25.6(0.9)   107.1(3.8)

The hydrogenation was carried out in *n*-hexane.

Rogers, D. W., Unpublished.

$C_{13}H_{26}$

(1)   1-Tridecene
        298 K   30.8(0.5)   128.7(2.0)

Hydrogenation was carried out in *n*-hexane. The value differs slightly from the original paper due to a revised standard $\Delta_{hyd}H$, Rogers, D. W., *J. Phys. Chem.* **1979**, *83*, 2430 (one page).

Rogers, D. W.; Skanupong, S. *J. Phys. Chem.* **1974**, *78*, 2569–2572.

$C_{14}H_{10}$

(1)   Diphenylacetylene
        298 K   64.1(1.1)   268.2(4.6)

$\Delta_{hyd}H$ was measured in ethanol and corrected for heat of solution. The value given is for a solid → solid reaction.

Flitcroft, T. L.; Skinner, H. A. *Trans. Faraday Soc.* **1958**, *54*, 47–53.

(2)   Diphenylacetylene
        298 K   59.6(0.6)   249.5(2.3)

The hydrogenation was carried out in *n*-hexane solution. The entry above is the arithmetic mean of two separate experiments consisting of 9 hydrogenation runs each.

Davis, H. E.; Allinger, N. L.; Rogers, D. W. *J. Org. Chem.* **1985**, *50*, 3601–3604.

$C_{14}H_{12}$

(1)   Diphenylethylene, (Z)
        cis-Stilbene
        302.1 K   25.8(0.2)   107.9(0.8)

$\Delta_{hyd}H$ is for the liquid → solution reaction (see note under 1-heptene). The solid sample was corrected to the liquid phase using the heat of fusion.

    355 K (gas phase, estimated)   26.3   110.0

Williams, R. B. *J. Amer. Chem. Soc.* **1942**, *64*, 1395–1404.

(2)   Diphenylethylene, (E)
           trans-Stilbene
           302.1 K   20.1(0.1)   84.3(0.4)

$\Delta_{hyd}H$ is for the liquid → solution reaction (see note under 1-heptene).

    355 K (gas phase, estimated)   20.6   86.2

Williams, R. B. *J. Amer. Chem. Soc.* **1942**, *64*, 1395–1404.

(3)   Heptafulvalene
           Cycloheptatrienylidenecycloheptatriene
           298 K   130.8(0.1)   547.1(0.6)

$\Delta_{hyd}H$ was measured in diethylcarbitol. Hydrogen uptake was for only 6 mols of H_2 so the result and the resonance energy derived from it (28 kcal mol^{-1}) should be regarded with some skepticism.

Turner, D. W.; Meador, W. R.; Doering, W. von E.; Knox, L. H.; Mayer, J. R.; Wiley, D. W. *J. Amer. Chem. Soc.* **1957**, *79*, 4127–4133.

(4)   Octalin
           Octalene
           1,2-Hexatrienylcyclooctatetraene
           298 K   157.2(0.3)   657.7(1.2)

$\Delta_{hyd}H$ was measured in cyclohexane. A correction was made of + 0.1 kcal mol^{-1} per double bond in the reactant for the heat of solution difference between reactant and product. Corrections were not made for the difference in heats of vaporization or sublimation in this study. A mixed reaction product was obtained and corrected to the major component using values of $\Delta_f H$ calculated by molecular mechanics. This is reflected in the relatively large experimental uncertainty.

Roth, W. R.; Lennartz, H.W.; Vogel, E.; Leiendecker, M.; Oda, M. *Chem. Ber.*, **1986**, *119*, 837–843.

$C_{14}H_{14}$

(1)   Dihydroheptafulvalene

298 K   138.8(0.2)   580.8(0.8)

$\Delta_{hyd}H$ was measured in acetic acid. Hydrogen uptake was normal. See heptafulvene.

Turner, D. W.; Meador, W. R.; Doering, W. von E.; Knox, L. H.; Mayer, J. R.; Wiley, D. W. *J. Amer. Chem. Soc.* **1957**, *79*, 4127–4133.

$C_{14}H_{18}$

(1)   1-Phenylcyclooctene
      298 K   21.3(0.2)   89.1(0.8)

Hydrogenation was carried out in *n*-hexane using styrene ($\Delta_{hyd}H$ = -28.20 kcal mol^{-1}) as a thermochemical standard. The tabulated value is the arithmetic mean of two separate experiments, each consisting of 9 runs of sample *vs.* standard. The uncertainty is estimated to express 95% confidence limits.

$C_{14}H_{24}$

(1)   Cyclotetradeca-1,8-diyne
      298 K   125.4(0.1)   524.7(0.6)

$\Delta_{hyd}H$ was measured in acetic acid solution. Solvent effects are significant for the alkynes and may cause the measured enthalpy changes to be up to 2.0 kcal mol^{-1} less exothermic than the gas-phase values.

Turner, R. B.; Jarrett, A. D.; Goebel, P.; Mallon, B. J. *J. Amer. Chem. Soc.* **1973**, *95*, 790–792.

(2)   1,1,4,4-Tetramethylcyclodec-6(7)-yne
      298 K   61.7   258.1

$\Delta_{hyd}H$ was measured in acetic acid solution. No experimental uncertainty was given.

Sicher, J.; Svoboda, M.; Zavada, J.; Turner, R. B.; Goebel, P. *Tetrahedron*, **1966**, *22*, 659–671.

(3)   1,1,4,4-Tetramethylcyclodec-7(8)-yne
      298 K   58.9   246.4

$\Delta_{hyd}H$ was measured in acetic acid solution. No experimental uncertainty was given.

Sicher, J.; Svoboda, M.; Zavada, J.; Turner, R. B.; Goebel, P. *Tetrahedron*, **1966**, *22*, 659–671.

$C_{14}H_{26}$

(1) 1,1,4,4-Tetramethylcyclodec-6(7)-ene, (Z)
cis-1,1,4,4-Tetramethylcyclodec-6(7)-ene
298 K   22.0   92.0

$\Delta_{hyd}H$ was measured in acetic acid solution. No experimental uncertainty was given. Results were applied to a conformational analysis of the 1,1,4,4-tetramethylcyclodecane system.

Sicher, J.; Svoboda, M.; Zavada, J.; Turner, R. B.; Goebel, P. *Tetrahedron*, **1966**, *22*, 659–671.

(2) 1,1,4,4-Tetramethylcyclodec-6(7)-ene, (E)
trans-1,1,4,4-Tetramethylcyclodec-6(7)-ene
298 K   25.4   106.3

$\Delta_{hyd}H$ was measured in acetic acid solution. No experimental uncertainty was given. Results were applied to a conformational analysis of the 1,1,4,4-tetramethylcyclodecane system.

Sicher, J.; Svoboda, M.; Zavada, J.; Turner, R. B.; Goebel, P. *Tetrahedron*, **1966**, *22*, 659–671.

(3) 1,1,4,4-Tetramethylcyclodec-7(8)-ene, (Z)
cis-1,1,4,4-Tetramethylcyclodec-7(8)-ene
298 K   28.7   120.0

$\Delta_{hyd}H$ was measured in acetic acid solution. No experimental uncertainty was given. Results were applied to a conformational analysis of the 1,1,4,4-tetramethylcyclodecane system.

Sicher, J.; Svoboda, M.; Zavada, J.; Turner, R. B.; Goebel, P. *Tetrahedron*, **1966**, *22*, 659–671.

(4) 1,1,4,4-Tetramethylcyclodec-7(8)-ene, (E)
trans-1,1,4,4-Tetramethylcyclodec-7(8)-ene
298 K   26.0   108.8

$\Delta_{hyd}H$ was measured in acetic acid solution. No experimental uncertainty was given. The unexpectedly large $\Delta_{hyd}H$ of 1,1,4,4-tetramethylcyclodec-7(8)-ene, (Z), which causes the usual E, Z stability relationship in cycloalkenes to be reversed, is brought about by strain. The conformation of the cyclodecane ring of 1,1,4,4-tetramethylcyclodec-7(8)-ene, (Z) forces a methyl group into the energetically unfavorable intraannular position as contrasted to the extraannular position it occupies in 1,1,4,4-tetramethylcyclodec-6(7)-ene, (E), 1,1,4,4-tetramethylcyclodec-6(7)-ene, (Z), and 1,1,4,4-tetramethylcyclodec-7(8)-ene, (E).

Sicher, J.; Svoboda, M.; Zavada, J.; Turner, R. B.; Goebel, P. *Tetrahedron*, **1966**, *22*, 659–671.

$C_{14}H_{28}$

(1)   1-Tetradecene
        298 K   30.5(0.4)   148.7(1.7)

Hydrogenation was carried out in *n*-hexane. The value differs slightly from the original paper due to a revised standard $\Delta_{hyd}H$, Rogers, D. W., *J. Phys. Chem.* **1979**, *83*, 2430 one page).

Rogers, D. W.; Skanupong, S. *J. Phys. Chem.* **1974**, *78*, 2569–2572.

(2)   3,4,-di(*t*-butyl)hex-3-ene, (E)
        298 K   38.1(0.3)   159.4(1.2)

The enthalpy of hydrogenation was corrected for small amounts of geometric rearrangement as detected in the product.

von E. Doering, W.; Roth, W. R.; Bauer, F.; Boenke, M.; Breuckmann, R.; Ebbrecht, M; Herbold, M.; Schmidt, R.; Lennartz, H.-W.; Lenoir, D.; Boese, R. *Chem. Ber.* **1989**, *122*, 1263–1275.

$C_{15}H_{24}$

(1)   2,2,5,5,8,8-Hexamethylnona-3,6-diyne
        298 K   130(1.0)   543.9(4.0)

The hydrogenation was carried out in *n*-hexane.

Scott, L. T.; Cooney, M. J.; Rogers, D. W.; Dejroongruang, K. *J. Amer. Chem. Soc.* **1988**, *110*, 7244–7245.

$C_{15}H_{30}$

(1)   1-Pentadecene
        298 K   30.4(0.2)   127.2(0.8)

Hydrogenation was carried out in *n*-hexane. The value differs slightly from the original paper due to a revised standard $\Delta_{hyd}H$, Rogers, D. W., *J. Phys. Chem.* **1979**, *83*, 2430 (one page).

Rogers, D. W.; Skanupong, S. *J. Phys. Chem.* **1974**, *78*, 2569–2572.

$C_{16}H_{10}$

(1) Diphenylbutadiyne
 298 K 118.5(1.0) 495.8(4.2)

The hydrogenation was carried out in *n*-hexane solution. Only the nonaromatic multiple bonds were hydrogenated. The entry above is the arithmetic mean of two separate experiments consisting of 9 hydrogenation runs each.

Davis, H. E.; Allinger, N. L.; Rogers, D. W. *J. Org. Chem.* **1985**, *50*, 3601–3604.

(2) Acepleiadylene
 298 88.4(0.5) 370.3(2.1)

Hydrogenation was carried out in glacial acetic acid over reduced Pt oxide. The reaction product was assumed to be tetrahydroacepleiadane (uptake of 5 mol equivalents of H_2 with retention of one benzenoid moiety) though this was not proven. The results were used to obtain a conjugation energy of approximately 19 kcal mol^{-1} for the three non-naphthalene double bonds in acepleiadylene. See also results for acepleyadane.

Turner, R. B.; Lindsay, W. S.; Boekelheide, V. *Tetrahedron* **1971**, *27*, 3341–3344.

$C_{16}H_{14}$

(1) 1,4-Diphenylbutadiene
 302.1 44.0(0.2) 184.3(0.8)

The value given is for the liquid → solution reaction (see note under 1-heptene).

 355 K (gas phase, estimated) 44.5 186.2

Williams, R. B. *J. Amer. Chem. Soc.* **1942**, *64*, 1395–1404.

$C_{16}H_{16}$

(1) Hexaquinacene
 C16 · Hexaquinacene
 298 K 71.2(1.0) 298(4)

The entry above is the arithmetic mean of three sets of experimental results, each of 9 determinations.

Rogers, D. W.; Loggins, S. A.; Samuel, S. D.; Finnerty, M. A.; Liebman, J. F. *Structural Chemistry* **1990**, *1*, 481–489.

(2) Acepleiadane
     298 K   29.7(0.1)   124.3(0.4)

Hydrogenation was carried out in glacial acetic acid over reduced Pt oxide. The reaction product was assumed to be tetrahydroacepleiadane (uptake of 5 mol equivalents of H_2 with retention of one benzenoid moiety) though this was not proven. The results were used to obtain a conjugation energy of approximately 25 kcal mol^{-1} for the two non-naphthalene double bonds in acepleiadane. See also results for acepleiadylene.

Turner, R. B.; Lindsay, W. S.; Boekelheide, V. *Tetrahedron* **1971**, *27*, 3341–3344.

$C_{16}H_{18}$

(1) Dihydrohexaquinacene
     298 K   47.3(1)   198(4)

The entry above is the arithmetic mean of three sets of experimental results, each of 9 determinations.

Rogers, D. W.; Loggins, S. A.; Samuel, S. D.; Finnerty, M. A.; Liebman, J. F. *Structural Chemistry* **1990**, *1*, 481–489.

$C_{16}H_{20}$

(1) Tetrahydrohexaquinacene
     298 K   24.8(0.6)   104 (3)

The entry above is the arithmetic mean of three sets of experimental results, each of 9 determinations.

Rogers, D. W.; Loggins, S. A.; Samuel, S. D.; Finnerty, M. A.; Liebman, J. F. *Structural Chemistry* **1990**, *1*, 481–489.

$C_{16}H_{28}$

(1) 1,1,4,4-Tetramethylcyclododec-8-yne
     298 K   61.3   256.5

$\Delta_{hyd}H$ was measured in acetic acid solution. No experimental uncertainty was given.

Sicher, J.; Svoboda, M.; Mallon, B. J.; Turner, R. B. *J. Chem. Soc. (B)*, **1968**, 441–447.

$C_{16}H_{30}$

(1)   1,1,4,4-Tetramethylcyclododec-8-ene, (Z)
        cis-1,1,4,4-Tetramethylcyclododec-8-ene
        298 K   31.2   130.3

$\Delta_{hyd}H$ was measured in acetic acid solution. No experimental uncertainty was given.

Sicher, J.; Svoboda, M.; Mallon, B. J.; Turner, R. B. *J. Chem. Soc. (B)*, **1968**, 441–447.

(2)   1,1,4,4-Tetramethylcyclododec-8-ene, (E)
        trans-1,1,4,4-Tetramethylcyclododec-8-ene
        298 K   26.6   111.3

$\Delta_{hyd}H$ was measured in acetic acid solution. No experimental uncertainty was given.

Sicher, J.; Svoboda, M.; Mallon, B. J.; Turner, R. B. *J. Chem. Soc. (B)*, **1968**, 441–447.

$C_{16}H_{32}$

(1)   1-Hexadecene
        298 K   30.3(0.4)   126.7(1.7)

The hydrogenation was carried out in *n*-hexane. The value differs slightly from the original paper due to a revised standard $\Delta_{hyd}H$, Rogers, D. W., *J. Phys. Chem.* **1979**, *83*, 2430 (one page).

Rogers, D. W.; Skanupong, S. *J. Phys. Chem.* **1974**, *78*, 2569–2572.

$C_{17}H_{12}$

(1)   1,1'-Spirobiindene
        298 K   61.0(1.0)   255.2(4.2)

Under the mild conditions used, only the nonaromatic double bonds were hydrogenated.

Hill, R. K.; Morton, G. H.; Rogers, D. W.; Choi, L. S. *J.Org. Chem.* **1980**, *45*, 5163–166.

$C_{17}H_{34}$

(1)   1-Heptadecene
       298 K   30.4(0.4)   127.0(1.7)

Hydrogenation was carried out in *n*-hexane. The value differs slightly from the original paper due to a revised standard $\Delta_{hyd}H$, Rogers, D. W., *J. Phys. Chem.* **1979**, *83*, 2430 (one page).

Rogers, D. W.; Skanupong, S. *J. Phys. Chem.* **1974**, *78*, 2569–2572.

$C_{18}H_{10}$

(1)   Angular[3]phenylene
       298 K   68.8(1)   288(4)

The hydrogenation was carried out in tetrahydrofuran against cyclohexene ($\Delta_{hyd}H$ = –28.44 kcal mol^{-1}) as a thermochemical standard, and was not corrected for solvent effects. The uncertainty estimate is the 95% confidence limit on a pooled data set of 26 individual enthalpy measurements in three separate experiments. See also C_3-symmetric [4]phenylene

Beckhaus, H.-D.; Faust, R.; Matzger, A. J.; Mohler, D. L.; Rogers, D. W.; Ruchardt, C.; Sawhney, A. K.; Verevkin, S. P.; Vollhardt, K. P. C.; Wolff, S. *J. Amer. Chem. Soc.* **2000**, *122*, 7819–7820.

$C_{18}H_{18}$

(1)   2,5-Diphenylhexa-1,5-diene
       298 K   56.6(0.2)   237.0(0.8)

The hydrogenation was carried out in isooctane and corrected for the double bond increment in heat of solution and for estimated differences in $\Delta_{hyd}H$ (see original publication, Table IV).

Roth, W. R.; Lennartz, H.-W.; Doering, W. von E.; Birladeanu, L; Guyton, C. A.; Kitagawa, T. *J. Amer. Chem. Soc.* **1990**, *112*, 1722–1732.

$C_{18}H_{36}$

(1)   1-Octadecene
       298 K   30.8(0.4)   128.8(1.7)

Hydrogenation was carried out in *n*-hexane. The value differs slightly from the original paper due to a revised standard $\Delta_{hyd}H$, Rogers, D. W., *J. Phys. Chem.* **1979**, *83*, 2430 (one page).

Rogers, D. W.; Skanupong, S. *J. Phys. Chem.* **1974**, *78*, 2569–2572.

$C_{19}H_{20}$

(1) 2,6-Diphenylhepta-1,6-diene
     298 K   55.7(0.2)   232.9(0.8)

The hydrogenation was carried out in isooctane and corrected for the double bond increment in heat of solution and for estimated differences in $\Delta_{hyd}H$ (see original publication, Table IV).

Roth, W. R.; Lennartz, H.-W.; Doering, W. von E.; Birladeanu, L; Guyton, C. A.; Kitagawa, T. *J. Amer. Chem. Soc.* **1990**, *112*, 1722–1732.

$C_{19}H_{38}$

(1) 1-Nonadecene
     298 K   30.7(0.4)   128.3(1.7)

The hydrogenation was carried out in *n*-hexane. The value differs slightly from the original paper due to a revised standard $\Delta_{hyd}H$, Rogers, D. W., *J. Phys. Chem.* **1979**, *83*, 2430 (one page).

Rogers, D. W.; Skanupong, S. *J. Phys. Chem.* **1974**, *78*, 2569–2572.

$C_{20}H_{30}$

(1) 2,2,5,5,8,8,11,11-Octamethyldodeca-3,6,9-triyne
     298 K   201(1.0)   878.6(4.0)

Hydrogenation was carried out in *n*-hexane.

Scott, L. T.; Cooney, M. J.; Rogers, D. W.; Dejroongruang, K. *J. Amer. Chem. Soc.* **1988**, *110*, 7244–7245.

(2) 2,4,6-Tri(t-butyl)bicyclo[3.3.0]octa-1,3,5,7-tetraene
     1,3,6-Tri(t-butyl)pentalene
     298 K   99.2(0.3)   415.1(1.3)

Roth, W. R.; Adamczak, O.; Breuckman, R.; Lennartz, H.-W.; Boese, R. *Chem. Ber.* **1991**, *124*, 2499–2521.

$C_{20}H_{40}$

(1)   1-Eicosene
      298 K   31.0(0.3)   129.9(1.3)

The hydrogenation was carried out in *n*-hexane. The value differs slightly from the original paper due to a revised standard $\Delta_{hyd}H$, Rogers, D. W., *J. Phys. Chem.* **1979**, *83*, 2430 (one page).

Rogers, D. W.; Skanupong, S. *J. Phys. Chem.* **1974**, *78*, 2569–2572.

$C_{24}H_{12}$

(1) C_3-symmetric[4]phenylene
      298 K   71.6(2)   300(9)

Hydrogenation was carried out in tetrahydrofuran against cyclohexene ($\Delta_{hyd}H = -28.44$ kcal mol^{-1}) as a thermochemical standard, and were not corrected for solvent effects. The uncertainty estimate is the 95% confidence limit on a pooled data set of 11 individual enthalpy measurements in five separate experiments. The central six membered ring is argued to be the theoretically significant planar *cyclohexatrien*e, that is, "localized" benzene.

Beckhaus, H.-D.; Faust, R.; Matzger, A. J.; Mohler, D. L.; Rogers, D. W.; Ruchardt, C.; Sawhney, A. K.; Verevkin, S. P.; Vollhardt, K. P. C.; Wolff, S. *J. Amer. Chem. Soc.* **2000**, *122*, 7819–7820.

$C_{25}H_{30}$

(1)   Decamethylcyclopentadeca-2,5,8,11,14-pentyne
      Decamethyl[5]pericycline
      298 K   341(2.2)   1427(9)

The hydrogenation was carried out in *n*-hexane.

Scott, L. T.; Cooney, M. J.; Rogers, D. W.; Dejroongruang, K. *J. Amer. Chem. Soc.* **1988**, *110*, 7244–7245.

$C_{25}H_{36}$

(1)   2,2,5,5,8,8,11,11,14,14-Decamethylpentadeca-3,6,9,12-tetrayne
      298 K   271(1)   1133 (4)

The hydrogenation was carried out in n-hexane.

Scott, L. T.; Cooney, M. J.; Rogers, D. W.; Dejroongruang, K. *J. Amer. Chem. Soc.* **1988**, *110*, 7244–7245.

$C_{26}H_{42}$

(1)   Cholest-1-ene
      298 K   27.3(0.2)   116.3(0.8)

$\Delta_{hyd}H$ was measured in acetic acid solution.

Turner, R. B.; Meador, W. R.; Winkler, R. E. *J. Amer. Chem. Soc.* **1957**, *79*, 4122–4127.

(2)   Cholest-2-ene
      298 K   25.8(0.2)   108.2(0.7)

$\Delta_{hyd}H$ was measured in acetic acid solution.

Turner, R. B.; Meador, W. R.; Winkler, R. E. *J. Amer. Chem. Soc..* **1957**, *79*, 4122–4127.

(3)   Cholest-3-ene
      298 K   28.0(0.2)   117.0(0.6)

$\Delta_{hyd}H$ was measured in acetic acid solution.

Turner, R. B.; Meador, W. R.; Winkler, R. E. *J. Amer. Chem. Soc..* **1957**, *79*, 4122–4127.

(4)   Cholest-5-ene
      298 K   25.8(0.1)   108.2(0.5)

$\Delta_{hyd}H$ was measured in acetic acid solution.

Turner, R. B.; Meador, W. R.; Winkler, R. E. *J. Amer. Chem. Soc..* **1957**, *79*, 4122–4127.

(5)   Cholest-6-ene
      298 K   27.4(0.1)   114.5(0.4)

$\Delta_{hyd}H$ was measured in acetic acid solution.

Turner, R. B.; Meador, W. R.; Winkler, R. E. *J. Amer. Chem. Soc.*. **1957**, *79*, 4122–4127.

$C_{29}H_{42}$

(1) 2,2,5,5,8,8,11,11,14,14,17,17-Dodecamethyloctadeca-3,6,9,12,15-pentayne
298 K   339(2.0)   1418(8)

The hydrogenation was carried out in *n*-hexane.

Scott, L. T.; Cooney, M. J.; Rogers, D. W.; Dejroongruang, K. *J. Amer. Chem. Soc.* **1988**, *110*, 7244–7245.

C_2H_4O

(1) Acetaldehyde
355 K   16.75(0.1)   70.1(0.4)

A small correction was made for incomplete reaction.

Dolliver, M. A.; Gresham, T. L.; Kistiakowsky, G. B.; Smith, E. A.; Vaughan, W. E. *J. Amer. Chem. Soc.* **1938**, *60*, 440–450.

C_3H_6O

(1) *n*-Propenol
     Allyl Alcohol
355 K   31.5(0.3)   132.1(1.3)

The main product was *n*-propanol, but some contamination with *n*-propyl ether was found which brought about a correction of 0.11 kcal mol^{-1}. The influence of the contaminant is included in the experimental uncertainty.

Dolliver, M. A.; Gresham, T. L.; Kistiakowsky, G. B.; Smith, E. A.; Vaughan, W. E. *J. Amer. Chem. Soc.* **1938**, *60*, 440–450.

(2) Acetone
355 K   13.4(0.1)   56.1(0.4)

A correction of 0.06 kcal mol^{-1} was made for incomplete reaction as detected by the presence of reactant in the product mix.

Dolliver, M. A.; Gresham, T. L.; Kistiakowsky, G. B.; Smith, E. A.; Vaughan, W. E. *J. Amer. Chem. Soc.* **1938**, *60*, 440–450.

C_4H_4O

(1)   Furan
        Oxacyclopentadiene
        355 K   36.6(0.1)   153.3(0.5)

Comparison with $2\times \Delta_{hyd}H$ (cyclopentene), $\Delta_{hyd}H$ (cyclopentadiene) or $\Delta_{hyd}H$ (ethyl vinyl ether) shows that the combination of two alternant double bonds and an alternant ether oxygen *in a ring* brings about a stabilization enthalpy of 15 – 20 kcal mol^{-1} relative to the three reference compounds given. Kistiakowsky refers to furan as "an almost wholly aromatic compound".

Dolliver, M. A.; Gresham, T. L.; Kistiakowsky, G. B.; Smith, E. A.; Vaughan, W. E. *J. Amer. Chem. Soc.* **1938**, *60*, 440–450.

C_4H_6O

(1)   2-Butenal
        Crotonaldehyde
        355 K   25.2(0.1)   105.8(0.4)

Substantial stabilization is brought about by the alternating arrangement of the C=C and C=O groups in crotonaldehyde.

Dolliver, M. A.; Gresham, T. L.; Kistiakowsky, G. B.; Smith, E. A.; Vaughan, W. E. *J. Amer. Chem. Soc.* **1938**, *60*, 440–450.

(2)   Divinyl ether
        355 K   57.2(0.1)   239.5(0.4)

Dolliver, M. A.; Gresham, T. L.; Kistiakowsky, G. B.; Smith, E. A.; Vaughan, W. E. *J. Amer. Chem. Soc.* **1938**, *60*, 440–450.

(3)   3-Oxacyclopentene
        298 K   25.6(0.3)   107.0(1.3)

Allinger, N. L.; Glasser, J. A.; Davis, H. E.; Rogers D. W. *J. Org. Chem.* **1981**, *46*, 658–661.

(4)   4-Oxacyclopentene
        298 K   28.0(0.3)   117.1(1.3)

Allinger, N. L.; Glasser, J. A.; Davis, H. E.; Rogers D. W. *J. Org. Chem.* **1981**, *46*, 658–661.

C_4H_8O

(1)   Ethyl vinyl ether
        355 K   26.7(0.1)   111.9(0.4)

The ethoxy group stabilizes the double bond in ethylene by about 6.1 kcal mol^{-1} relative to hydrogen.

Dolliver, M. A.; Gresham, T. L.; Kistiakowsky, G. B.; Smith, E. A.; Vaughan, W. E. *J. Amer. Chem. Soc.* **1938**, *60*, 440–450.

(2)   Ethyl vinyl ether
        298 K   26.5(0.2)   110.9(0.8)

Allinger, N. L.; Glasser, J. A.; Davis, H. E.; Rogers D. W. *J. Org. Chem.* **1981**, *46*, 658–661.

(3)   Methyl ethyl ketone
        355 K   13.2(0.1)   55.2(0.4)

A correction of 0.06 kcal mol^{-1} was made for incomplete reaction as detected by the presence of reactant in the product mix.

Dolliver, M. A.; Gresham, T. L.; Kistiakowsky, G. B.; Smith, E. A.; Vaughan, W. E. *J. Amer. Chem. Soc.* **1938**, *60*, 440–450.

C_5H_6O

(1)   2-Cyclopenten-1-one
        298 K   23.2(0.6)   97.1(2.7)

Hydrogenation was carried out in a mixed solvent of tetrahydrofuran and *n*-hexane using cyclohexene or 1-hexene ($\Delta_{hyd}H$ = –118.8 or –126.6 kJ mol^{-1}) as a thermochemical standard. The results were corrected for solvent effects in the mixed solvent used as a calorimeter fluid. The tabulated value is the arithmetic mean of two separate experiments, each consisting of 9 runs of sample *vs.* standard. The uncertainty is estimated to express 95% confidence limits.

Rogers, D. W.; Zhao, Y.; Traetteberg, M.; Hulce, M.; Liebman, J. F. *J. Chem. Thermodynam.* **1998**, *30*, 1393–1400.

C_5H_8O

(1)   Cyclopentanone
      355 K   12.5(0.2)   52.3(0.8)

A correction of 0.12 kcal mol^{-1} was made for incomplete reaction as detected by the presence of reactant in the product mix. $\Delta_{hyd}H$ is less than half that of a monoalkene.

Conn, J. B.; Kistiakowsky, G. B.; Smith, E. A. *J. Amer. Chem. Soc.* **1939**, *61*, 1868–1876.

(2)   Dihydropyran
      2H-Pyran-3,4-dihydro
      3-Oxacyclohexene
      298 K   24.7(0.3)   103.3(1.3)

Allinger, N. L.; Glasser, J. A.; Davis, H. E.; Rogers D. W. *J. Org. Chem.* **1981**, *46*, 658–661.

$C_5H_{10}O$

(1)   2-Ethoxypropene
      355 K   25.1(0.1)   105.0(0.4)

The oxygen atom is capable of conjugating with double bonds in either the C=C-C=O or C=C-C-O- configuration. The resulting stabilization energy is slightly smaller than that of a carbon-carbon double bond.

Dolliver, M. A.; Gresham, T. L.; Kistiakowsky, G. B.; Smith, E. A.; Vaughan, W. E. *J. Amer. Chem. Soc.* **1938**, *60*, 440–450.

      2-Methoxybut-2-ene
      355 K   24.8(0.2)   103.8(0.8)

See 2-ethoxypropene.

Dolliver, M. A.; Gresham, T. L.; Kistiakowsky, G. B.; Smith, E. A.; Vaughan, W. E. *J. Amer. Chem. Soc.* **1938**, *60*, 440–450.

C_6H_8O

(1)   2-Cyclohexen-1-one
      298 K   26.1(0.6)   109.2(2.9)

Hydrogenation was carried out in a mixed solvent of tetrahydrofuran and *n*-hexane using cyclohexene or 1-hexene ($\Delta_{hyd}H$ = –118.8 or –126.6 kJ mol^{-1}) as a thermochemical standard. The results were corrected for solvent effects in the mixed solvent used as a calorimeter fluid. The tabulated value is the arithmetic mean of two separate experiments, each consisting of 9 runs of sample vs. standard. The uncertainty is estimated to express 95% confidence limits.

Rogers, D. W.; Zhao, Y.; Traetteberg, M.; Hulce, M.; Liebman, J. F. *J. Chem. Thermodynam.* **1998**, *30*, 1393–1400.

$C_6H_{10}O$

(1)   Cyclohexanone
        355 K   15.4(0.1)   64.5(0.6)

A correction of 0.10 kcal mol^{-1} was made for incomplete reaction as detected by the presence of reactant in the product mix. $\Delta_{hyd}H$ is approximately half that of a monoalkene.

Conn, J. B.; Kistiakowsky, G. B.; Smith, E. A. *J. Amer. Chem. Soc.* **1939**, *61*, 1868–876.

C_7H_6O

(1)   Tropone
        1,3,5-Cycloheptatriene-7-one
        Heptatrienylium oxide
        298 K   67.6(0.3)   282.8(1.3)

$\Delta_{hyd}H$ was measured in acetic acid and corrected for $\Delta_{sol'n}H$ of tropone in acetic acid using Pd on BaSO$_4$ catalyst. The reaction product was cycloheptanone. The resonance stabilization by the carbonyl group is only slightly less than that of the methylene group in heptafulvene. Compare tropylidene and heptafulvene.

Turner, D. W.; Meador, W. R.; Doering, W. von E.; Knox, L. H.; Mayer, J. R.; Wiley, D. W. *J. Amer. Chem. Soc.* **1957**, *79*, 4127–4133.

$C_7H_{12}O$

(1)   5-Methylhex-5-en-2-one
        298 K   27.1(0.6)   113.2(2.7)

Hydrogenation was carried out in a mixed solvent of tetrahydrofuran and *n*-hexane using cyclohexene or 1-hexene ($\Delta_{hyd}H = -118.8$ or -126.6 kJ mol^{-1}) as a thermochemical standard. The results were corrected for solvent effects in the mixed solvent used as a calorimeter fluid. The tabulated value is the arithmetic mean of two separate experiments, each consisting of 9 runs of sample *vs.* standard. The uncertainty is estimated to express 95% confidence limits.

Rogers, D. W.; Zhao, Y.; Traetteberg, M.; Hulce, M.; Liebman, J. F. *J. Chem. Thermodynam.* **1998**, *30*, 1393–1400.

$C_8H_{12}O$

(1) 9-Oxabicyclo[3.3.1]non-1-ene
     Cyclooct-1-ene oxide
     298 K   37.0(0.2)   154.7(0.8)

Allinger, N. L.; Glasser, J. A.; Davis, H. E.; Rogers D. W. *J. Org. Chem.* **1981**, *46*, 658–661.

$C_3H_4O_2$

(1) Acrylic acid
     298 K   30.3(0.2)   127.0(0.8)

$\Delta_{hyd}H$ was measured in ethanol and corrected by a separate heat of solution measurement. The value given is for the liquid → liquid reaction.

Skinner, H. A.; Snelson A. *Trans. Faraday Soc.* **1959**, *55*, 404–407.

$C_4H_4O_2$

(1) Methyl propiolate
     Propynoic acid, methyl ester
     Propargilic acid, methyl ester
     298 K   79.1(1.1)   331(4.4)

$\Delta_{hyd}H$ was measured in ethanol and corrected for the heat of solution. The value given is for a liquid → liquid reaction. Hydrogen uptake was slightly under theoretical indicating the possibility of 2-4% methanol impurity.

Flitcroft, T. L.; Skinner, H. A. *Trans. Faraday Soc.* **1958**, *54*, 47–53.

(2) Butynoic acid
     298 K   72.4(1.1)   303(4.4)

$\Delta_{hyd}H$ was measured in ethanol and corrected by a separate heat of solution measurement experiment. The value given is for a (liquid) → (liquid) reaction.

Flitcroft, T. L.; Skinner, H. A. *Trans. Faraday Soc.* **1958**, *54*, 47–53.

$C_4H_6O_2$

(1) Vinylacetate
316 K   31.1(0.1)   130.2(0.4)

Dolliver, M. A.; Gresham, T. L.; Kistiakowsky, G. B.; Smith, E. A.; Vaughan, W. E. *J. Amer. Chem. Soc.* **1938**, *60*, 440–450.

The carboxyl oxygen does not enter into conjugative stabilization interaction with the double bond.

(2) Methacrylic acid
298 K   28.2(0.3)   118.0 (1.3)

$\Delta_{hyd}H$ was measured in ethanol and corrected by a separate heat of solution measurement. The value given is for the liquid → liquid reaction. Given the α methyl stabilization in methacrylic acid, there is no evidence for stabilization by the carboxyl oxygen.

Skinner, H. A.; Snelson A. *Trans. Faraday Soc.* **1959**, *55*, 404–407.

$C_5H_6O_2$

(1) Pentynoic acid
298 K   65.8(1.2)   275(5.0)

$\Delta_{hyd}H$ was measured in ethanol and corrected by a separate heat of solution measurement experiment. The value given is for a solid → solid reaction.

Flitcroft, T. L.; Skinner, H. A. *Trans. Faraday Soc.* **1958**, *54*, 47–3.

$C_5H_8O_2$

(1) Methylmethacrylate
Methylpropenoic acid methyl ester
355 K   28.6(0.1)   119.8(0.4)

The result is consistent with stabilization by the methyl group α to the double bond and no energetic interaction with the oxygen atoms.

Dolliver, M. A.; Gresham, T. L.; Kistiakowsky, G. B.; Smith, E. A.; Vaughan, W. E. *J. Amer. Chem. Soc.* **1938**, *60*, 440–450.

$C_6H_{10}O_2$

(1) 3,4-Dihydro-2-methoxy-2H-pyran
298 K   25.9(0.3)   108.3(1.3)

The methyl group opposite to the double bond is slightly destabilizing.

Allinger, N. L.; Glasser, J. A.; Davis, H. E.; Rogers D. W. *J. Org. Chem.* **1981**, *46*, 658–661.

(2) 5,6-Dihydro-4-methoxy-2H-pyran
298 K   23.1(0.2)   96.7(0.8)

The methyl group α to the double bond is slightly stabilizing. The compound showed slight turbidity during sample preparation.

Allinger, N. L.; Glasser, J. A.; Davis, H. E.; Rogers D. W. *J. Org. Chem.* **1981**, *46*, 658–661.

$C_7H_{10}O_2$

(1) Methylhex-2-ynoate
298 K   71.1(1.2)   297.5(5.0)

$\Delta_{hyd}H$ was measured in ethanol and corrected by a separate heat of solution measurement experiment. The value given is for a liquid → liquid reaction. Compare 1-pentyne.

Flitcroft, T. L.; Skinner, H. A. *Trans. Faraday Soc.* **1958**, *54*, 47–53.

(2) 3-Carbomethoxy-1,2-dimethylcyclopropene
1,2-Dimethyl-3-carbomethoxycyclopropene
298 K   40.2(0.3)   173.3(1.3)

$\Delta_{hyd}H$ was measured in acetic acid. Hydrogen was absorbed in excess of the theoretical amount. The value given is permol of H_2. $\Delta_{hyd}H$ is influenced by strain energy in both the cyclopropene ring of the reactant and the *cis*-cyclopropane ring of the product.

Turner, R. B.; Goebel, P.; Mallon, B. J.; von E. Doering, W.; Coburn Jr., J. F.; Pomeranz, M. *J. Amer. Chem. Soc.* **1968**, *90*, 4315–4322.

$C_8H_{14}O_2$

(1) 1,4-bis(vinyloxy)butane
298 K   53.0(0.3)   221.9(1.3)

Compare ethyl vinyl ether and bis(vinyloxy)ethyl ether. No significant stabilizing effects are found here, in contrast to the vinylic-oxygen interaction in divinyl ether.

Allinger, N. L.; Glasser, J. A.; Davis, H. E.; Rogers D. W. *J. Org. Chem.* **1981**, *46*, 658–661.

$C_9H_6O_2$

(1) Phenylpropionic acid
     298 K   72.6(1.1)   303.7(4.6)

$\Delta_{hyd}H$ was measured in ethanol and corrected for heat of solution. There is an evident destabilizing effect of the phenyl group and the carboxyl group on acetylene. Its magnitude is not clear because $\Delta_{hyd}H$ is open to some question owing to scarcity of sample which precluded more than two runs. The value given is for a solid → solid reaction. Compare phenylacetylene.

Flitcroft, T. L.; Skinner, H. A. *Trans. Faraday Soc.* **1958**, *54*, 47–53.

$C_{10}H_{10}O_2$

(1) Methyl cinnamate
     2-Propenoic acid, 3-phenyl-, methyl ester
     302 K   28.2(0.2)   117.9(1.0)

$\Delta_{hyd}H$ is given for the liquid → solution reaction (see note under 1-heptene). Solid sample was corrected to the liquid phase using the heat of fusion.

     355 K (gas phase, estimated)   28.7

Williams, R. B. *J. Amer. Chem. Soc.* **1942**, *64*, 1395–1404.

(2) Methyl *trans*-cinnamate
     3-phenyl-2-Propenoic acid, methyl ester, (E)
     302 K   24.2(0.1)   101.2(0.4)

$\Delta_{hyd}H$ is given for the liquid → solution reaction (see note under 1-heptene).

     355 K (gas phase, estimated)   24.7   103.3

Williams, R. B. *J. Amer. Chem. Soc.* **1942**, *64*, 1395–1404.

$C_{11}H_{18}O_2$

(1) 2-Methyl-3-isobutoxycyclohex-2-enone
     298 K   25.4(0.7)   106.1(3.0)

Hydrogenation was carried out in a mixed solvent of tetrahydrofuran and n-hexane using cyclohexene or 1-hexene ($\Delta_{hyd}H$ = –118.8 or –126.6 kJ mol^{-1}) as a thermochemical standard. The results were corrected for solvent effects in the mixed solvent used as a calorimeter fluid. The tabulated value is the arithmetic mean of two separate experiments, each consisting of 9 runs of sample *vs.* standard. The uncertainty is estimated to express 95% confidence limits.

Rogers, D. W.; Zhao, Y.; Traetteberg, M.; Hulce, M.; Liebman, J. F. *J. Chem. Thermodynam.* **1998**, *30*, 1393–1400.

$C_{16}H_{30}O_2$

(1) Hexadec-9-enoic acid, (Z)
     cis-9-Hexadecenoic acid
     Palmitoleic acid

     298 K   29.9(0.2)   125.1(1.0)

Hydrogenation was carried out in n-hexane solution which approximates the non-interactive environment of the gas phase.

Rogers, D. W.; Hoyte, O. P. A; Ho, R. K. C. *J. Chem. Soc., Faraday 1* **1978**, *74*, 46–52.

$C_{17}H_{32}O_2$

(1) Methyl hexadec-9-enoate, (Z)
     Hexadec-9-enoic acid methyl ester, (Z)
     cis-9-Hexadecenoic acid methyl ester
     Methyl Palmitoleate
     298 K   30.0(0.3)   125.5(1.3)

Hydrogenation was carried out in n-hexane solution which approximates the non-interactive environment of the gas phase. Value differs slightly from the original paper due to a revised standard $\Delta_{hyd}H$, Rogers, D. W., *J. Phys. Chem.* **1979**, *83*, 2430 (one page).

Rogers, D. W.; Siddiqui, N. A. *J. Phys. Chem.* **1975**, *79*, 574–577.

(2) Methyl hexadec-9-enoate, (E)
     Hexadec-9-enoic acid methyl ester, (E)

trans-9-Hexadecenoic acid methyl ester
Methyl Palmitelaidate
298 K   33.1(0.6)   138.5(2.5)

Hydrogenation was carried out in *n*-hexane. A limited amount of sample was available. The result is probably about 3 kcal mol^{-1} too exothermic. Value differs slightly from the original paper due to a revised standard $\Delta_{hyd}H$, Rogers, D. W., *J. Phys. Chem.* **1979**, *83*, 2430 (one page).

Rogers, D. W.; Siddiqui, N. A. *J. Phys. Chem.* **1975**, *79*, 574–577.

$C_{18}H_{30}O_2$

(1)   Octadeca-9,12,15-trienoic acid, (Z,Z,Z)
        cis-9-cis-12-cis-15-Octadecatrienoic acid
        Linolenic Acid
        298 K   90.9(0.5)   380.2(1.9)

Hydrogenation was carried out in *n*-hexane solution which approximates the non-interactive environment of the gas phase.

Rogers, D. W.; Hoyte, O. P. A; Ho, R. K. C. *J. Chem. Soc., Faraday 1* **1978**, *74*, 46–52.

$C_{18}H_{32}O_2$

(1)   Octadeca-9,12-dienoic acid, (Z,Z)
        cis-9-cis-12-Octadienoic acid
        Linoleic Acid
        298 K   61.3(0.4)   254.4(1.5)

Hydrogenation was carried out in *n*-hexane solution which approximates the non-interactive environment of the gas phase.

Rogers, D. W.; Hoyte, O. P. A; Ho, R. K. C. *J. Chem. Soc., Faraday 1* **1978**, *74*, 46–52.

(2)   Octadeca-9,12-dienoic Acid, (E,E)
        trans-9-trans-12-Octadecadienoic acid
        Linoelaidic Acid
        298 K   59.5(0.1)   248.8(0.5)

Hydrogenation was carried out in *n*-hexane solution which approximates the non-interactive environment of the gas phase.

Rogers, D. W.; Hoyte, O. P. A; Ho, R. K. C. *J. Chem. Soc., Faraday 1* **1978**, *74*, 46–52.

$C_{18}H_{34}O_2$

(1)   Octadec-9-enoic acid, (Z)
        cis-9-Octadecenoic acid
        Oleic Acid
        298 K   29.7(0.3)   124.4(1.2)

The result is the arithmetic mean of two complete experiments. Hydrogenation was carried out in n-hexane solution which approximates the non-interactive environment of the gas phase.

Rogers, D. W.; Hoyte, O. P. A; Ho, R. K. C. *J. Chem. Soc., Faraday 1* **1978**, *74*, 46–52.

(2)   Octadec-9-enoic Acid, (E)
        trans-9-Octadecenoic acid
        Elaidic Acid
        298 K   28.7(0.5)   120.2(2.0)

Hydrogenation was carried out in n-hexane solution which approximates the non-interactive environment of the gas phase.

Rogers, D. W.; Hoyte, O. P. A; Ho, R. K. C. *J. Chem. Soc., Faraday 1* **1978**, *74*, 46–52.

$C_{19}H_{32}O_2$

(1)   Methyl octadeca-9,12,15-trienoate, (Z)
        Octadeca-9,12,15-trienoic acid methyl ester, (Z)
        cis-9-cis-12-cis-15-Octadecatrienoic acid methyl ester
        Methyl linolenate
        298 K   86.10.6)   360(2.5)

Hydrogenation was carried out in n-hexane solution which approximates the non-interactive environment of the gas phase. Value differs slightly from the original paper due to a revised standard $\Delta_{hyd}H$, Rogers, D. W., *J. Phys. Chem.* **1979**, *83*, 2430 (one page).

Rogers, D. W.; Siddiqui, N. A. *J. Phys. Chem.* **1975**, *79*, 574–577.

$C_{19}H_{34}O_2$

(1)   Methyl octadeca-9,12-dienoate, (Z,Z)
        Octadeca-9,12-dienoic acid methyl ester, (Z,Z)
        cis-9-cis-12-Octadecadienoic acid methyl ester

Methyl linoleate
298 K   59.3(0.4)   248(1.7)

Hydrogenation was carried out in *n*-hexane solution which approximates the non-interactive environment of the gas phase. Value differs slightly from the original paper due to a revised standard $\Delta_{hyd}H$, Rogers, D. W., *J. Phys. Chem.* **1979**, *83*, 2430 (one page).

Rogers, D. W.; Siddiqui, N. A. *J. Phys. Chem.* **1975**, *79*, 574–577.

(2)   Methyl octadeca-9,12-dienoate, (E,E)
        Octadec-9,12-enoic acid methyl ester, (E,E)
        trans-9-*trans*-12-0ctadecadienoic acid methyl ester
        Methyl linoelaidate
        298 K   56.4(0.1)   236(0.4)

Hydrogenation was carried out in *n*-hexane solution, which approximates the non-interactive environment of the gas phase. Value differs slightly from the original paper due to a revised standard $\Delta_{hyd}H$, Rogers, D. W., *J. Phys. Chem.* **1979**, *83*, 2430 (one page).

Rogers, D. W.; Siddiqui, N. A. *J. Phys. Chem.* **1975**, *79*, 574–577.

$C_{19}H_{36}O_2$

(1)   Methyloctadec-9-enoate, (Z)
        Octadec-9-enoic acid methyl ester, (Z)
        cis-9-Octadecenoic acid methyl ester
        Methyl Oleate
        298 K   30.0(0.3)   125.5(1.3)

Hydrogenation was carried out in very dilute *n*-hexane solution which approximates the non-interactive environment of the gas phase. Value differs slightly from the original paper due to a revised standard $\Delta_{hyd}H$, Rogers, D. W., *J. Phys. Chem.* **1979**, *83*, 2430 (one page).

Rogers, D. W.; Siddiqui, N. A. *J. Phys. Chem.* **1975**, *79*, 574–577.

(2)   Methyloctadec-9-enoate, (E)
        Octadec-9-enoic acid methyl ester, (E)
        trans-9-Octadecenoic acid methyl ester
        Methyl elaidate
        298 K   29.0(0.2)   121.3(0.8)

Hydrogenation was carried out in very dilute n-hexane solution which approximates the non-interactive environment of the gas phase. Value differs slightly from the original paper due to a revised standard $\Delta_{hyd}H$, Rogers, D. W., *J. Phys. Chem.* **1979**, *83*, 2430 (one page).

Rogers, D. W.; Siddiqui, N. A. *J. Phys. Chem.* **1975**, *79*, 574–577.

$C_6H_{10}O_3$

(1)   2,5-Dimethoxy-2,5-dihydrofuran
        2,5-Dihydro-2,5-dimethoxyfuran
        298 K   31.4(0.2)   131.3(0.8)

Allinger, N. L.; Glasser, J. A.; Davis, H. E.; Rogers D. W. *J. Org. Chem.* **1981**, *46*, 658–661.

$C_8H_{14}O_3$

(1)   bis[(Vinyloxy)ethyl]ether
        298 K   53.7(0.4)   224.7(1.8)

Allinger, N. L.; Glasser, J. A.; Davis, H. E.; Rogers D. W. *J. Org. Chem.* **1981**, *46*, 658–661.

$C_{24}H_{38}O_3$

        3-alpha-Hydroxy-Δ^{11}-cholenic acid
        Lithocholenic acid
        298 K   28.9(0.2)   121.0(0.8)

$\Delta_{hyd}H$ was measured in acetic acid solution.

Turner, R. B.; Meador, W. R.; Winkler, R. E. *J. Amer. Chem. Soc.* **1957**, *79*, 4122–4127.

$C_4H_2O_4$

(1)   Acetylene dicarboxylic acid·2H$_2$O
        298 K   83.1(1.2)   348.9(5.0)

$\Delta_{hyd}H$ was measured in ethanol and corrected for the heat of solution of the product, succinic acid. Measurements were carried out on both the crystalline dihydrate and the anhydride(below). The heat of solution correction is complicated by the heat of solution, in ethanol, of water present in the dihydrate. Some question was raised as to whether residual water remained in the "anhydrous" solid.

Flitcroft, T. L.; Skinner, H. A. *Trans. Faraday Soc.* **1958**, *54*, 47–53.

(1')   Acetylene dicarboxylic acid
         298 K   86.8(1.1)   363.2(4.6)

$\Delta_{hyd}H$ was measured in ethanol and corrected for the heat of solution of the product, succinic acid. Measurements were carried out on both the crystalline dihydrate (above) and the anhydride. The heat of solution correction is complicated by the heat of solution, in ethanol, of water present in the dihydrate. Some question was raised as to whether residual water remained in the "anhydrous" solid.

Flitcroft, T. L.; Skinner, H. A. *Trans. Faraday Soc.* **1958**, *54*, 47–53.

$C_4H_4O_4$

(1)   Maleic acid
         cis-Butenedioic acid
         Butenedioic acid, (Z)
         298 K   31.3(0.3)   130.9(1.3)

$\Delta_{hyd}H$ was measured in ethanol, and is uncorrected for the heat of solution of succinic acid. The value given is for a solid → solid reaction.

Flitcroft, T. L,; Skinner, H. A.; Whiting, M. C. *Trans. Faraday Soc.* **1957**, *53*, 784–790.

(1')   Maleic acid
         cis-Butenedioic acid
         Butenedioic acid, (Z)
         298 K   36.6(0.4)   153.1(1.7)

The solution value was corrected for enthalpies of solution. The enthalpy of solution of succinic acid (product) is 5.3(0.1) kcal mol^{-1}, accounting for the large difference between this entry and the previous one. The value given is for a solid → solid reaction.

Flitcroft, T. L,; Skinner, H. A.; Whiting, M. C. *Trans. Faraday Soc.* **1957**, *53*, 784–790.

(1")   Maleic acid
         cis-Butendioic acid
         Butenedioic acid, (Z)
         298 K   36.3(0.2)   151.9(0.4)

$\Delta_{hyd}H$ was measured in ethanol. The value given is for the solid → solid reaction.

Skinner, H. A.; Snelson A. *Trans. Faraday Soc.* **1959**, *55*, 404–407.

(2) Fumaric Acid
     trans-Butenedioic acid
     Butenedioic acid, (E)
     298 K   25.8(0.1)   107.9(0.4)

$\Delta_{hyd}H$ was measured in ethanol. The value given is for the solid → solution reaction.

Flitcroft, T. L,; Skinner, H. A.; Whiting, M. C. *Trans. Faraday Soc.* **1957**, *53*, 784–790.

(2') Fumaric acid
      trans-Butendioic acid
      Butenedioic acid, (E)
      298 K   31.2(0.3)   130.3(1.3)

The solution value was corrected for the enthalpy of solution. The value given is for a solid → solid reaction. The enthalpy of solution of succinic acid (product) is 5.3(0.1) kcal mol^{-1}, accounting for the large difference between this entry and the previous one.

Flitcroft, T. L,; Skinner, H. A.; Whiting, M. C. *Trans. Faraday Soc.* **1957**, *53*, 784–790.

$C_8H_{12}O_4$

(1) Diethyl maleate
     Butenedioic acid-(Z) diethyl ester
     302   33.5(0.2)   140.2(0.8)
     355 K (gas phase, estimate) 34.0   142

Williams notes an "exaltation" of $\Delta_{hyd}H$ by the carboxyl group, as did Kistiakowsky (see vinyl acetate).

Williams, R. B. *J. Amer. Chem. Soc.* **1942**, *64*, 1395–1404.

(2) Diethyl fumarate
     Butenedioic acid-(E) diethyl ester
     302   29.3(0.2)   122.6(0.8)
     355 (gas phase-estimate)   29.8   125

Williams, R. B. *J. Amer. Chem. Soc.* **1942**, *64*, 1395–1404.

(3) Diethyl fumarate
     298 K   28.9(0.1)   121.1(0.4)

$\Delta_{hyd}H$ was measured in acetic acid solution.

Turner, R. B.; Meador, W. R.; Winkler, R. E. *J. Amer. Chem. Soc.*. **1957**, *79*, 4116–4121.

$C_{57}H_{92}O_6$

(1) Glycerol octadeca-9,12,15-trienoate, (Z,Z,Z)
     Trilinolenin
     298 K   270.7(0.6)   1132(2.3)

Hydrogenation was carried out in *n*-hexane solution, which approximates the non-interactive environment of the gas phase. $\Delta_{hyd}H$ is relative to 125.1 kcal mol^{-1}, (exothermic) for methyl oleate, which was taken as the standard value.

Rogers; D. W.; Choudhury, D. N. *J. Chem. Soc., Faraday 1* **1978**, *74*, 2868–2872.

$C_{57}H_{98}O_6$

(1) Glycerol octadeca-9,12-dienoate, (Z,Z)
     Trilinolein
     298 K   180.9(0.3)   757(1.6)

Hydrogenation was carried out in very dilute *n*-hexane solution which approximates the non-interactive environment of the gas phase. $\Delta_{hyd}H$ is relative to 125.1 kcal mol^{-1} (exothermic) for methyl oleate which was taken as the standard value.

Rogers; D. W.; Choudhury, D. N. *J. Chem. Soc., Faraday 1* **1978**, *74*, 2868–2872.

$C_{57}H_{104}O_6$

(1) Glycerol octadeca-9-enoate, (Z)
     Triolein
     298 K   90.9(0.8)   381(3.4)

Hydrogenation was carried out in *n*-hexane solution which approximates the non-interactive environment of the gas phase. $\Delta_{hyd}H$ is relative to 125.1 kcal mol^{-1} (exothermic) for methyl oleate, which was taken as the standard value.

Rogers; D. W.; Choudhury, D. N. *J. Chem. Soc., Faraday 1* **1978**, *74*, 2868–2872.

(2) Glycerol octadeca-9-enoate, (E)
     Trielaidin
     298 K   90.1(0.5)   377(2.0)

Hydrogenation was carried out in n-hexane solution which approximates the noninteractive environment of the gas phase. $\Delta_{hyd}H$ is relative to 125.1 kcal mol^{-1} (exothermic) for methyl oleate, which was taken as the standard value.

Rogers; D. W.; Choudhury, D. N. *J. Chem. Soc., Faraday 1* **1978**, *74*, 2868–2872.

(3)   Glycerol octadec-6-enoate, (Z)
       Tripetroselenin
       298 K   90.5(0.2)   378.5(0.7)

Hydrogenation was carried out in n-hexane solution which approximates the noninteractive environment of the gas phase. $\Delta_{hyd}H$ is relative to 125.1 kcal mol^{-1} (exothermic) for methyl oleate which was taken as the standard value.

Rogers; D. W.; Choudhury, D. N. *J. Chem. Soc., Faraday 1* **1978**, *74*, 2868–2872.

$C_7H_{15}N_3$

(1)   3-Ethyl-3-azidopentane
       298 K   80.6(2.0)   337(8.4)

$\Delta_{hyd}H$ was measured in n-hexane solution. Hydrogenation to the amine and nitrogen is thought to be clean and quantitative. Interference was observed, ascribable to adsorption of the amine product at the beginning, and catalyst poisoning at the end of the experiment.

Wayne, G. S.; Snyder, G. J.; Rogers, D. W. *J. Amer. Chem. Soc.* **1993**, *115*, 9860–9861.

$C_8H_{11}N_3$

(1)   3-Phenyl-3-azidopropane
       298 K   82.3(1.4)   344(5.9)

$\Delta_{hyd}H$ was measured in n-hexane solution. Hydrogenation to the amine and nitrogen is thought to be clean and quantitative. Interference was observed, ascribable to adsorption of the amine product at the beginning, and catalyst poisoning at the end of the experiment.

Wayne, G. S.; Snyder, G. J.; Rogers, D. W. *J. Amer. Chem. Soc.* **1993**, *115*, 9860–9861.

$C_{10}H_{15}N_3$

(1)   1-Azidoadamantane
       298 K   80.3(1.6)   336(6.7)

$\Delta_{hyd}H$ was measured in *n*-hexane solution. Hydrogenation to the amine and nitrogen is thought to be clean and quantitative. Interference was observed, ascribable to adsorption of the amine product at the beginning, and catalyst poisoning at the end of the experiment.

Wayne, G. S.; Snyder, G. J.; Rogers, D. W. *J. Amer. Chem. Soc.* **1993**, *115*, 9860–9861.

C_6H_9NO

(1)   Hex-2-ynoamide
          298 K   72.9(0.8)   305.0(3.2)

$\Delta_{hyd}H$ was measured in ethanol. The value given is for a solid → solid reaction corrected for $\Delta_{sol'n}H$.

Flitcroft, T. L.; Skinner, H. A. *Trans. Faraday Soc.* **1958**, *54*, 47–53.

$C_5H_6F_2$

(1)   1-Vinyl-2,2-difluorocyclopropane
          298 K   79.7(0.2)   333.5(0.8)

The reaction was carried out in isooctane and the result was corrected for the difference in heats of solution of the reactant and product. The reaction product was 2,2-difluoropentane.

Roth, W. R.; Kirmse, W.; Hoffmann, W.; Lennartz, H.-W. *Chem. Ber.* **1982**, *115*, 2508–2515.

(2)   3,3-Difluorocyclopentene
          298 K   29.0(0.1)   121.3(0.4)

The reaction was carried out in isooctane. The reaction product was difluorocyclopentane. Compare $\Delta_{hyd}H$ = –26.8(0.1) kcal mol^{-1} for hydrogenation of cyclopentene to cyclopentane (Roth, W. R.; Lennartz, H.-W. *Chem. Ber.* **1980**, *113*, 1806–1817).

Roth, W. R.; Kirmse, W.; Hoffmann, W.; Lennartz, H.-W. *Chem. Ber.* **1982**, *115*, 2508–2515.

$C_6H_8F_2$

(1)   2,2-Difluorocyclopropyl-1-propene
          298 K   76.8(0.3)   321.3(1.2)

The reaction was carried out in isooctane and the result was corrected for the difference in heats of solution of the reactant and product. The reaction product was 2,2-difluorohexane.

Roth, W. R.; Kirmse, W.; Hoffmann, W.; Lennartz, H.-W. *Chem. Ber.* **1982**, *115*, 2508–2515.

(2)   1-Vinyl-2,2-difluoro-3-methylcyclopropane
      298 K   76.6(0.1)   320.5(0.4)

The reaction was carried out in isooctane. The reaction product was 3,3-difluorohexane.

Roth, W. R.; Kirmse, W.; Hoffmann, W.; Lennartz, H.-W. *Chem. Ber.* **1982**, *115*, 2508–2515.

$C_7H_4F_2$

(1)   7,7-Difluorobenzocyclopropene

      298 K   132.0   552

Numerical data in Table 3 do not agree with descriptive material in the discussion. It is difficult to decide from this paper what the value is or its reliability. This datum is not recommended.

Roth, W. R.; Figge, H-J. *European Journal of Organic Chemistry* **2000**, 1983–1986.

C_7H_7Cl

(1)   Cycloheptatrienylium chloride
      Tropylium chloride
      298 K   86.2(0.4)   361(1.7)

The result is calculated taking into account several interfering reactions and some impurity in the sample. It is calculated on the basis of hydrogen absorbed. The reactant is unstable in air. Both this and the following reaction amount to hydrogenation of the tropylium ion and a solvated proton. The heat of formation of dissolved tropylium chloride is 19.8 kcal mol^{-1}.

Turner, R. B.; Prinzbach. H.; von E. Doering, W. *J. Amer. Chem Soc.* **1960**, *82*, 3451–3454.

C₇H₇Br

(1)   Cycloheptatrienylium bromide
      Tropylium bromide
      298 K   89.1(0.2)   373(0.8)

The result is calculated on the basis of hydrogen absorbed. Both this and the preceding reaction amount to hydrogenation of the tropylium ion and a solvated proton. The discrepancy is ascribed to incomplete dissociation of the ions involved.

Turner, R. B.; Prinzbach. H.; von E. Doering, W. *J. Amer. Chem Soc.* **1960**, *82*, 3451–3454.

---

a) Some enthalpies of hydrogenation of alkyl halides to produce hydrogen halides have been measured. They are not included here. (See Jensen (1976), also Conn, J. B.; Kistiakowsky, G. B.; Smith, E. A. *J. Amer. Chem. Soc.* **1938**, *60*, 2764–2771.)

b) A few common names are included in this table.

c) It is a longstanding tradition in thermochemistry to report results to the nearest 0.01 kcal mol^{-1} even though statistical treatment may not support this many significant figures. Except for a few of the smaller compounds for which $\Delta_{hyd}H$ is known with high precision, most results are given with one significant figure beyond the decimal point. The numbers in parentheses after selected values are experimental uncertainties given as our best estimate of 95% confidence limits. Often these uncertainties are larger than the values given for the uncertainty in the original paper, reflecting a certain degree of conservatism on the part of the author.

Chapter 3
Computational Thermochemistry

3.1 Introduction to Computational Thermochemistry

Many potentially interesting and useful hydrogenation experiments are not carried out because they are impractical owing to reactant polymerization, or inadvisable because they are dangerous, as in the case of polyalkynes, some of which detonate spontaneously. The enthalpy of hydrogenation of a simple monoalkene

$$C_nH_{2n} + H_2 \rightarrow C_nH_{2n+2}$$

is the difference between the enthalpies of formation of the alkane product and the alkene reactant

$$\Delta_{hyd}H°_{298} = \Delta_f H°_{298}(C_nH_{2n+2}) - \Delta_f H°_{298}(C_nH_{2n}) \quad (3.1)$$

because $\Delta_f H°_{298}(H_2) = 0$ by definition. If we could *calculate* the enthalpy of formation in the standard state $\Delta_f H°_{298}$ of both the reactants and products for a hydrogenation reaction, we would have the $\Delta_{hyd}H°_{298}$, also in the standard state. Analogous equations would hold for polyalkenes, alkynes and polyalkynes in which the degree of reactant unsaturation is greater than it is for C_nH_{2n} and the number of moles of H_2 consumed is greater than one.

For most of the 20th century, chemists struggled to solve the intractable equations of molecular quantum mechanics or to tease out the many modes of classical motion executed by bound atoms in order to calculate the thermodynamic properties of molecules. In the 1970s successes were achieved in classical mechanics, most notably by Allinger with his molecular mechanics (MM) series of computer programs, and in quantum mechanics by Pople (Nobel prize 1998) with his series of Gaussian quantum mechanics programs.

In the meantime, Benson had developed an additive approach to the thermochemistry of molecules, based on the idea that thermodynamic properties like $\Delta_f H_{298}$ can, at least to a certain extent, be regarded as the sum of $\Delta_f H_{298}$ values ascribed to constituent parts of the molecule, such as the C–C bond or the –CH$_2$– group. These constituent values he called *bond additivity values* or *group additivity values*. We shall see the distinction below. Although the objective of these calculations is the standard state enthalpy of formation, superscript ° will not be used in the notation because calculated $\Delta_f H_{298}$ values are approximate by definition.

Dewar, on the other hand, retained the mathematical structure of a purely quantum mechanical treatment at the beginning of his attempts to calculate molecular properties but when faced with daunting integrals, as he often was, he replaced them by empirically determined parameters. He did not do this to make things easy for himself, indeed, some of the procedures for determining parameters are themselves daunting, rather he did it to save calculation time and computer storage capacity. The method developed by Dewar and others is called a *semi-empirical* approach, being between the purely quantum mechanical method of Pople (called *ab initio*) and the fully empirical methods of Allinger and Benson. Computers are used in all four methods above. We shall examine all of these approaches to calculating $\Delta_f H_{298}$ in this chapter.

There is no right or wrong method in computational molecular thermochemistry. Each method is appropriate in its own context. The trade-off among the four methods above is speed *vs.* quantum mechanical rigor. The order of methods from fastest to slowest is: *group additivity, molecular mechanics, semi-empirical,* and *ab initio*. Quantum mechanical rigor is in the reverse order. The faster methods are used to solve problems involving larger molecules, which are computer intensive. The rigorous methods are used to study small differences in structure and energy among relatively small molecules, ions, or molecular species of fleeting existence such as free radicals.

Regrettably, the spirit of free exchange of computer programs practiced by pioneers in this field no longer prevails. Programs to calculate molecular structure and energy can be prohibitively expensive for individuals or small science departments. Therefore we shall describe some public domain software that allows one to calculate $\Delta_f H_{298}$ and $\Delta_{hyd} H_{298}$ of simple molecules at no expense beyond the basic microcomputer operating system and an internet connection. For small research projects, a graphical user interface (gui) is recommended at modest cost. For full scale research, the reader may want to apply for a grant of computer time at one of the National Science Foundation Supercomputing Centers, for example, see *allocations.ncsa.uiuc.edu*.

3.2 Molecular Modeling

As the name implies, molecular modeling includes simple construction of a three dimensional structure with wooden balls and dowels to depict atoms and bonds, or any other attempt to construct a visual aid that facilitates thinking about the invisible world of the molecule. In contemporary usage, however, the term denotes a more abstract model of molecular structure, usually a mathematical model.

The most fundamental molecular property obtained from a molecular model is its geometry. Geometry is expressed either in the form of Cartesian or internal coordinates. Figure 3.1 shows the output of a molecular modeling program in the form of Cartesian coordinates for the water molecule. Each triplet x, y, and z is the position vector of the atom identified in the first column on the left. The position vectors are given in vertical columns from left to right. For this parameter file, the molecule lies nearly in the x–y plane because the z component of each atom is small. Notice that the difference between the H and O coordinates in the 1,1 and 2,1 positions of the geometry matrix is about 0.96 Å. Compare this with the experimental value of 0.96 Å for the H – O bond distance in H_2O (Eğe, 1994). The coordinates in Fig. 3.1 form a 3×3 matrix. In general, the matrix is $n \times 3$ for a molecule with n atoms.

```
H  -0.857070   0.748629   0.093156
O   0.089540   0.753627   0.119839
H   0.338853  -0.159618   0.145072
```

Fig. 3.1 Cartesian coordinates for the H$_2$O molecule. Units are Å.

An expression of the geometry of molecules in *internal coordinates* that is widely used is the z-matrix. A z-matrix representation of the geometry of the water molecule is given in Fig. 3.2.

```
H   0.000000     0.000000   0.000000
O   0.947000     0.000000   0.000000
H   0.947000   104.999992   0.000000
```

Fig. 3.2 Internal coordinates for the H$_2$O molecule.

In z-matrix format, a guide atom is chosen as the reference point. In this case, the reference atom is one of the hydrogens H. It is given the coordinates H   0.000000   0.000000   0.000000 in the top row of Fig 3.2. A second atom (in this case O) is located 0.947 Å from the reference atom. The third atom, H is located 0.947 Å from the second atom. The three atoms make a simple angle of 104.999992° (i.e., 105° within computer bit limits). Since any three points determine a plane, the dihedral angle of the third atom with the other two atoms is 0°, given in the third column of Fig. 3.2. This procedure of locating atoms by distance and angle can be continued indefinitely. In general, dihedral angles will not be 0°.

For larger molecules one may wish to express simple and dihedral angles relative to some atom or atoms other than the first atom taken. A more general form of the z-matrix is given in Fig. 3.3. Here the O atom is chosen (arbitrarily) as the reference point. The integer following each atom designator other than the first gives the atom number to which a distance or an angle is referred. For example, H in the second line, is at a distance of R2 from atom 1, that is, it is 0.947000 Å from the oxygen atom. In the third line, R3 is measured relative to atom 1 and the simple angle A3 is measured with atom 2 as the vertex. Continuing with a dihedral angle A-B-C-D would require specification of atom D for measurement of the dihedral angle that A makes with D as seen down the B-C axis. Input distances and angles may not be known in a research problem but an informed guess can always be made. The computational procedures given below carry out iterative improvement on an approximate input geometry.

```
O
H, 1, R2
H, 1, R3, 2, A3
Variables
R2=0.947000
R3=0.947000
A3=105.000000
```

Fig. 3.3 A z-matrix for the H$_2$O molecule.

Many variations on z-matrix input files are possible. For example, Fig. 3.3 can be rewritten in a more compact form by recognizing that the variables R2 and R3 are redundant. There are some arbitrary format requirements peculiar to each program hence there is a learning curve for each one. One compact variation on Fig. 3 is shown in Fig. 3.3a. In this system, the blank line 4 is essential.

```
o
h 1 r
h 1 r a

r 0.947
a 105.0
```

Fig. 3.3a A variant on the z-matrix for H$_2$O.

Molecular properties related to geometry, such as dipole moments, are usually calculated, as are the common thermodynamic functions. Having the model (usually computer output) in mathematical form does not preclude, of course, use of the mathematical coordinates to draw two- or three-dimensional diagrams (*visualization*) on the computer screen to appeal to the pictorial thinking favored by most chemists. Indeed the drawing may be made to mimic the familiar wooden ball-and-stick models, with the difference that we expect it to be more accurate in its geometric features than a wooden model.

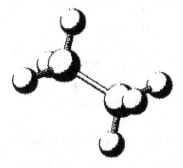

Fig. 3.4 Visualization of the output for the ethane molecule (PCMODEL).

If the enthalpies of formation can be obtained by any or all of the computational methods above for a hydrogenation reaction of known $\Delta_{hyd}H_{298}$, we have a check on both the calculational method and the experimental procedure used to measure $\Delta_{hyd}H_{298}$. Having two, or even several ways to calculate or measure the same enthalpies does not imply wasted effort in the form of redundancy. Results of calculations vary in their degree of rigor and the completeness of their parameter sets. Experimental results, even from highly reputable research groups can be wrong. New computational methods appear regularly and old ones are continually updated. New experimental results appear regularly and are incorporated into existing parameter sets. The objective, of course, after having verified a computational method by successful cross checking with known results, is to use it to obtain results that are not known. In the context of contemporary thermochemistry, this usually means obtaining unknown $\Delta_f H_{298}$ values pertaining to molecules, ions or radicals that are not amenable to experiment.

3.3 Additivity Methods

An additivity relationship involves some molecular property that is the sum of contributions from smaller units within the molecule. For example, the mass of a molecule is, to a very good approximation, the sum of the masses of the atoms that comprise it. (This relationship is not exact because of relativistic effects.)

3.3.1 Bond additivity

Bond-energy schemes in their simplest form involve the assumption that each bond in a molecule contributes a partial $\Delta_f H_{298}$(bond) to the total $\Delta_f H_{298}$(molecule)

Table 3.1 Average Bond Energies (Eğe, 1994)

Bond	kcal mol^{-1}	kJ mol^{-1}
H-H	104	435
H-C	99	414
C-C	83	347
C=C	146	611
C≡C	200	837

In the simple hydrogenation of ethene

$$CH_2=CH_2 + H\text{-}H \rightarrow CH_3\text{-}CH_3 \qquad (3.2)$$

the total energy of bonds broken (Table 3.1) is $146 + 104 = 250$ kcal mol^{-1} and the total energy of bonds formed is $2(99) + 83 = 281$ kcal mol^{-1}. Bond breaking requires energy input to the system therefore the energy necessary to break bonds during the reaction is positive. Bond formation releases energy from the system and is negative. The total energy change of the hydrogenation reaction is $250 + (-281) = -31$ kcal mol^{-1} which is not a bad first approximation to the experimental value of -32.6 kcal mol^{-1} (Chap. 2).

Before we become too pleased with this result, we should ask why there is any error at all. After all, really good experimental results have an uncertainty of 0.1-0.2 kcal mol^{-1}, about an order of magnitude smaller than the discrepancy between the calculated $\Delta_{hyd}H_{298}$ and the experimental value. Also, working from different sources soon reveals disagreement as to the values of the bond energies themselves, especially for multiple bonds such as C=C and C≡C. Aren't the molecules in reaction (3.2) held together by a known number of chemical bonds with known energies? The answer is that they are indeed held together by a known number of chemical bonds but that the nature of the chemical bond in all its aspects, including bond energy, is somewhat dependent on environment.

This is easily seen in the stepwise dissociation of the four equivalent bonds in CH_4.

$$CH_4 \rightarrow CH_3 + H \qquad \Delta H = 432 \text{ kJ mol}^{-1}$$

$$CH_3 \rightarrow CH_2 + H \qquad \Delta H = 471$$

$$CH_2 \rightarrow CH + H \qquad \Delta H = 422$$

$$CH \rightarrow C + H \qquad \underline{\Delta H = 339}$$

$$\Delta H(\text{total}) = 1664$$

(3.3)

These results show clearly that bond energies, of which C–H is an example, are not equal to the bond dissociation energy of any single C–H linkage in methane. Although the bonds in CH_4 are truly equivalent at the outset, after a hydrogen has been taken away, we are dealing with an entirely different reaction. The reaction

$$CH_4 \rightarrow CH_3 + H$$

in the sequence 3.3 is not the same as

$$CH_3 \rightarrow CH_2 + H$$

hence the energy changes shouldn't be the same either. Rather the *average* dissociation energy is taken as the bond energy

$$\Delta H_{298}(\text{C-H}) = \frac{1664}{4} = 416 \text{ kJ mol}^{-1}.$$

Even this value is only "close" to the tabulated value of 414 kJ mol^{-1} for the C–H bond energy in Table 3.1. The remaining small difference arises because bond energies are fitted parameters. Total energies or enthalpies in addition to those of methane, for example those of higher alkanes, are used in the fitting or averaging procedure. This also explains why different bond energies may be found in different sources. Parameter fitting is a subjective process which can involve different choices of the database used to generate the parameters.

In the unique case of diatomic molecules, however, bond energies and dissociation energies are, in principle, identical and exact (although, there may be disagreement in the literature as to what the exact values are). The enthalpy change for the reaction

$$H_2 \rightarrow 2 H \qquad \Delta H° = 436.0 \text{ kJ mol}^{-1}$$

which is the bond dissociation energy of H$_2$, is frequently quoted (CRC Handbook of Chemistry and Physics) as 104.2 kcal mol^{-1} of H$_2$ gas or 52.10 kcal mol^{-1} of H atoms produced. In the case of diatomic molecules like H$_2$, but unlike CH$_4$, there are no second, third, *etc.* dissociation energies to be averaged in, hence the enthalpy or energy of dissociation is the same as the bond energy. Notice that the terms bond *energy* and bond *enthalpy* are used interchangeably in much of the descriptive literature. Certainly differences between energy and enthalpy are thermodynamically real and can be quite significant in some applications. They are, however, negligible in comparison to the error bought about by averaging and database selection necessary to approximate bond contributions to the energy or the enthalpy change of a hydrocarbon reaction.

Yet another important concept comes to light when we examine the seemingly simple method of approximating $\Delta_f H_{298}$ of a molecule using bond energies. When we write a bond energy with appropriate caveats regarding its approximate nature, we say something like, "The energy of the C–H bond is 432 kJ mol^{-1}" for the first step in the sequence 3.3. A legitimate question is: Relative to what?

Looking again at the successive bond breaking reactions of CH_4, it is evident that the total energy of bond breaking for all bonds in the sequence is relative to atoms *in the gaseous state*, $C(g) + 4H(g)$, not to the elements in the standard state as is true of normal enthalpies of formation. We can think of the bond breaking reactions 3.3 in reverse as energies of formation from high energy components, the atoms in the gaseous state. (It takes a lot of energy to vaporize graphite.) This is legitimate because the standard state against which we measure energies is an arbitrary choice. We are free to change it as the occasion demands, but we must be consistent within any single application.

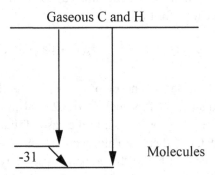

Fig. 3.5 Formation of ethane and ethane from the gaseous atoms. Formation of ethene is on the left and ethane is on the right. Not to scale.

The same kind of formation reaction can be written for any hydrocarbon molecule or, indeed for any molecule, given sufficient energy information. Thus both ethene and ethane in the hydrogenation reaction (3.2) can be thought of as being formed by combining gaseous hydrogen atoms with gaseous carbon atoms to form a stable molecule. That no such reaction takes place in the laboratory has no influence on the energetics of the situation. As long as a thermodynamic cycle can be drawn in which all energies save one are known or can be measured, the unknown energy can be calculated by difference. The experimental enthalpies of atomization of both hydrogen and carbon have been measured but the experiment is difficult and somewhat prone to error leading to slight discrepancies among literature values.

3.3.2 Group additivity

Group additivity schemes are based on the idea that molecular properties are the aggregate of contributions from identifiable *groups*, such as $-CH_3$.

Table 3.2 Selected group additivity values for alkanes (from Cohen and Benson, 1993). The *gauche* and 1-5 corrections arise from environmental factors.

Group	kcal mol^{-1}	kJ mol^{-1}
C-(C)(H)$_3$ = P	-10.00	-41.8
C-(C)$_2$(H)$_2$ = S	-5.00	-20.9
C-C(C)$_3$(H) = T	-2.40	-10.0
C-(C)$_4$ = Q	-0.10	-0.42
Gauche	0.80	3.35
1-5	1.60	6.69

The top entry in Table 3.2 is for a carbon C connected to another carbon (C) and three hydrogens (H). It is also designated P, indicating that it is a "primary" carbon atom. The other three carbon atoms in the table are denoted similarly as secondary, tertiary and quaternary, S, T, and Q. When two primary carbon atoms (two methyl groups) join to form an ethane molecule, CH$_3$–CH$_3$, the sum of group enthalpies from Table 3.1 is -20.00 kcal mol^{-1} (experimental value $\Delta_f H_{298} = -20.1 \pm 0.1$ kcal mol^{-1}. [Statistical note: while it would be improper to report an experimental measurement of -20.1 ± 0.1 as -20.00, one is free to select a *parameter* -20.00 from the range given by -20.0 ± 0.1.] Remember that molecules other than *n*-propane will be used in the averaging and parameterizing procedure. Whether the choice of a parameter is judicious or injudicious would then be determined by some statistical measure of agreement between a set of predictions made using the selected parameter and experimental measurements covering the same data set. The standard deviation or mean absolute deviation (e.g. Rogers, 2003) would be an appropriate statistical measure of agreement.

Table 3.3 Selected group additivity values for alkenes (from Cohen and Benson, 1993)

Bond	kcal mol^{-1}	kJ mol^{-1}
C_d-$(H)_2$	6.27	26.23
C_d-$(C)(H)$	8.55	35.77
C_d-$(C)_2$	10.19	42.63
C_d-$(C_d)H$	6.78	28.37
C_d-$(C_d)(C)$	8.76	36.65
C-$(C_d)(H)_3$	−10.00	−41.84

To predict $\Delta_{hyd}H_{298}$(ethene), we shall also need $\Delta_f H_{298}$(ethene) which can be estimated as the sum of two $=CH_2$ contributions taken from the table of alkene group contributions, Table 3.3. In Table 3.3, carbon atoms designated with a subscripted $_d$ are double bonded carbons, so $=CH_2$ is given as $C_d(H)_2$. The predicted $\Delta_f H_{298}$(ethene) is $2(6.27) = 12.54$ kcal mol^{-1} which leads to the predicted $\Delta_{hyd}H_{298}$(ethene)

$$\Delta_{hyd}H_{298}(\text{ethene}) = -20.00 - 12.54 = -32.54 \text{ kcal mol}^{-1}$$

as compared to the experimental value of -32.58 ± 0.05 kcal mol^{-1} (Chap. 2). The advantages of simplicity and accuracy are evident in this example. This is, however, an unusually favorable case; normally the uncertainty in $\Delta_f H_{298}$ is 0.1–0.2 kcal mol^{-1} and the "average error" (Cohen and Benson, 1993) of the group additivity method relative to experimental values is 0.4–0.6 kcal mol^{-1}.

In a like way, we can obtain a group enthalpy parameter for the $-CH_2-$ group in propane. The equation

$$\Delta_f H_{298}(n\text{-propane}) = 2\Delta_f H_{298}(-CH_3) + \Delta_f H_{298}(-CH_2-)$$

expresses the enthalpy of formation of *n*-propane. The experimental value of $\Delta_f H_{298}$(*n*-propane) = -25.0 ± 0.1 kcal mol^{-1} leads to $\Delta_f H_{298}(-CH_2-) = -25.00 - (-20.00) = -5.00$ kcal mol^{-1}. On the assumption that group enthalpy parameters remain the same from one molecule to the next, we can predict $\Delta_f H_{298}$ for any of the *n*-alkanes by adding 2[$\Delta_f H_{298}(-CH_3)$] to the requisite number of $\Delta_f H_{298}(-CH_2-)$ group enthalpies. Branched alkanes require new parameters to account for the degree of branching. Alkenes, alkynes and molecules containing one or more heteroatoms can be treated in the same way, but each extension requires one or more new parameters.

Entries 5 and 6 in Table 3.2 labeled *gauche* and 1-5, relate to interactions between groups that are not bound to the same atom. For example, both terminal $-CH_3$ groups in propane are bound to the central $-CH_2-$ but the terminal $-CH_3$ groups in *n*-butane are not bound to a common atom; one is bound to one $-CH_2-$ and the other terminal $-CH_3$ group is bound to a different $-CH_2-$ group. These terminal $-CH_3$ groups interfere with each other more in the *gauche* conformation than they do in the *trans* conformation. *Gauche n*-butane is destabilized relative to *trans*. The destabilization is small (0.80 kcal mol^{-1}) and positive. The *gauche* interaction is a "1-4" interaction because the interacting groups are numbered 1 and 4 as we count down the *n*-butane chain. There is also a 1-5 interaction energy included in Table 3.2.

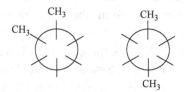

Fig. 3.6 *Gauche* and *trans n*-butane looking down the 2-3 bond axis. The *gauche* conformation is on the left.

So far we have assumed that we know all $\Delta_f H_{298}$(*n*-alkane) values necessary to obtain group or bond parameters. Once having generated a sufficient number of parameters from a small but reliable database, we can, in principle, calculate any desired $\Delta_f H_{298}$. Given the nearly 20 million molecules that have been characterized over the history of chemistry, one can see the strength of purely empirical methods. With such an enormous amount of thermochemical information yet to be determined, simplicity is more than a strength, in many applications it is a necessity.

On the other hand, the last two entries in Table 3.2 hint at the weakness of empirical methods. They show that the "environment" of a group may lead to many parameters other than simple group energies when we are talking about a molecule of, say, biochemical or medical complexity. For every new group and every new environmental factor characterized, one or more new parameters must be determined, perhaps causing the parameter base to grow beyond reasonable bounds. Of course, any environmental factor that is real but is *not* included in the parameter set, causes an error in the calculated thermochemical properties.

3.3.3 *The thermochemical database*

Up to now, we have assumed that we have a database large enough to calculate all the group energy values we need. Clearly that database has to come from somewhere. In fact, most of the enthalpies of formation we have (*webbook.nist.gov*) come from combustion studies. Just about all organic compounds burn. Many of them burn quantitatively to form only CO_2 and H_2O. If a compound burns quantitatively, its enthalpy of formation can be determined from the heat given off, its *heat of combustion*.

We have seen that $\Delta_f H^\circ_{298}$ is the enthalpy change of a compound associated with its formation, in the standard state, from elements also in the standard state, In some cases, $\Delta_f H^\circ_{298}$ can be measured directly. For example $\Delta_f H^\circ_{298}(CO_2)$ is determined by burning graphite in oxygen.

$$C(graphite) + O_2(g) \rightarrow CO_2(g)$$

This reaction can be carried out under conditions that permit the temperature change of the surroundings to be measured and so to permit determination of the amount of heat given off per mole of $CO_2(g)$ formed. [Although the actual combustion takes place at a high temperature, conditions can be controlled so that the system starts at $T = 298.15$ K and returns to a temperature that is above 298.15 by a very small ΔT. This difference causes an error in enthalpy that is negligible by comparison to errors from other sources.]

Although flow calorimetric experiments have been carried out at constant P which yield $\Delta_f H_{298}$ directly (for example, Wiberg and Fenoglio, 1968), combustion is usually carried out at constant volume in a closed container. Closed container or *bomb calorimetric* measurements of ΔT yield the *energy* of formation in the standard state, for example, $\Delta_f E^\circ_{298}(CO_2)$. This datum leads to $\Delta_f H^\circ_{298}(CO_2)$ through the defining equation

$$\Delta H = \Delta E + P\Delta V$$

at constant P (Klotz and Rosenberg, 2000).

The same kind of experiment can be carried out on C, H_2, C_2H_4, and C_2H_6. Combustion of the two hydrocarbons, C_2H_4 and C_2H_6, yields CO_2 and H_2O. Using the results of all these combustion experiments, thermochemical cycles can be set up with only $\Delta_f H^\circ_{298}$ (hydrocarbon) as the unknown quantity. After calculating $\Delta_f H^\circ_{298}$ for the two hydrocarbons in Eq. (3.2), we can obtain $\Delta_{hyd} H^\circ_{298}$ (alkene), *having used a sequence of reactions that is entirely independent of any direct experimental measurement of* $\Delta_{hyd} H_{298}$. The two methods then serve as a cross check on one another. Combustion thermochemistry generates the database from which virtually all of the empirical parameters used in computational thermochemistry are drawn and it provides experimental standards against which new computational methods are tested.

Determining $\Delta_f H^\circ_{298}$ (hydrocarbons) by combustion, despite its seeming simplicity, is not an easy job. Combustion thermochemistry requires meticulous control of experimental conditions (Steele et al., 2002). The difficulty of this task and its expense in time and money mean that researchers in need of $\Delta_f H^\circ_{298}$ data are likely to find vast gaps in the thermochemical record. At present, chemists are synthesizing new compounds far more rapidly than thermochemical properties are being measured, hence the reliance in many contemporary studies and in industrial laboratories on computational methods. The simplest of these is undoubtedly the group additive method of Benson. Despite the undeniable popularity of group-enthalpy schemes in practical problem solving, one soon becomes aware of their lack of theoretical underpinning. Also, like other methods parameterized on ground state energies, they are mainly used to predict ground state energies.

In the next section, we shall take a first step in the direction of mathematical rigor by describing molecular mechanics, a computational method that establishes not only the ground state equilibrium geometry of a molecule, along with properties like dipole moment that are dependent on the geometry, but also gives an estimate of $\Delta_f H_{298}$. Molecular mechanics is based on the classical equations of motion (harmonic oscillation, torsional oscillation, *etc.*) but its force constants are determined at least partly by empirical fitting procedures using $\Delta_f H_{298}$ values drawn from classical thermochemistry.

3.4 Molecular Mechanics

The classical equations of motion used in molecular mechanics (MM) are only slightly more difficult to solve than simple additive bond energy equations hence MM calculations are fast and not very demanding of computer resources. In molecular mechanics, one determines the structure of a molecule from a knowledge of the *force field*, a collection of empirical force constants governing, in principle, all classical mechanical interatomic interactions within the molecule. In practice, it is not feasible for a parameter set to include all possible interactions within a complicated molecule. One hopes that all *significant* interactions have been included in the force field.

A molecular structure is arrived at in MM by minimizing the total potential energy arising from all classical modes of motion included in the program. Structures arrived at in MM are usually a compromise involving distortion of some bonds and angles to satisfy the geometric requirements of others. For example, atoms A, B, and C might be connected in the form of a bent molecule such that there is an equilibrium bond distance A–B, an equilibrium bond distance A–C, and equilibrium simple bond angle ABC. If now, a bond is formed between atoms A and C, such that a *triangular* molecule ABC is formed, the A–C bond distance will not, in all likelihood, be just right to accommodate the bond angle ABC. All bonds and angles will be distorted a little to form the new molecular structure and the equilibrium bond energies of all bonds and angles in the bent triatomic molecule will be driven up to accommodate the new structural requirements. This increment contributes to the internal energy of the molecule.

In more complicated molecules, there will be many distortions made to accommodate the equilibrium molecular structure. Each distortion has a potential energy associated with it. The sum of all distortion potential energies is called the total *steric energy* of the molecule. If this steric energy is added to a sum of "normal" bond enthalpies, the result is $\Delta_f H_{298}$ for the molecule. Entropies and heat capacities are also calculated by molecular mechanics, yielding Gibbs free energies and equilibrium constants (Allinger *et al.*, 1996).

Especially for determining $\Delta_f H_{298}$ of alkenes and alkanes, existing force fields and "normal" bond enthalpy parameters are very good. Molecular mechanics estimates of $\Delta_f H_{298}$, for 20 randomly chosen simple alkanes, alkenes, alcohols, amines, both cyclic and acyclic, have a mean unsigned difference from experimental results of 0.2 kcal mol^{-1} = 0.8 kJ mol^{-1}. The principal drawback of molecular mechanics in calculating $\Delta_f H_{298}$ is that it is a parameterized method. Either force constants or bond energy parameters may be missing for the enthalpy one wishes to calculate.

The principle of classical equation solving, which is at the heart of MM, can be appreciated by imagining two objects, connected by a spring, executing simple harmonic motion collinear with the spring. If the force constant k of the spring is known, it is possible to calculate the potential energy V of the system at any separation x of the objects as $V = kx^2 / 2$.

For any nonequilibrium distance, the potential energy will be higher than it is at the equilibrium separation. Thus, one way to determine the equilibrium length of the spring is to calculate the energy of the system for many arbitrary separations and throw out all but the lowest energy solution. If enough calculations are done at sufficiently fine discrimination, the equilibrium separation, hence the resting length of the spring can be approximated to any desired accuracy. If there are many objects connected by many springs, there will be an overall equilibrium configuration that leads to an overall potential energy minimum. In general this configuration will not be at the rest lengths of any of the springs. The equilibrium configuration will be a compromise of all forces acting on all objects.

In MM, the "objects" in the last paragraph are atoms and the springs are chemical interactions, including the chemical bonds that hold the molecule together along with other repulsive and attractive interactions within the molecule. At the potential energy minimum, the position of each atom will be as near to the bottom of its parabolic potential well as it can be without undue distortion of the forces acting on all the other atoms.

The actual MM program operates on an input geometry in the form of a set of Cartesian x, y, z coordinates of each atom in the molecule (see Fig. 3.1). The MM program changes the coordinates in a systematic way, keeping changes that bring about a decrease in the energy of the molecule as a whole and rejecting geometry changes that increase the energy. This is continued until the energy does not change for a small change in geometry. In most cases, a minimum has been reached in the total energy of the molecule. The geometry at which this minimum is found is the equilibrium geometry, which can be presented by the program as a new Cartesian coordinate set or in pictorial form (see Fig. 3.4) Since the energy was minimized to obtain the overall equilibrium structure, both energy and structure emerge from the same procedure.

The requirement that the energy change be at or near zero for a small change in geometry is the mathematical condition for an extremum. If the initial estimate of the geometry was reasonably close to the equilibrium geometry, the extremum will be a minimum and the true equilibrium geometry will have been reached. If not, the extremum may be a minimum relative to its immediate environment but not the lowest or *global* minimum. It may also be a saddle point or, rarely, a maximum.

For large molecules or molecules with numerous close lying minima, a good starting geometry may be hard to find (Burkert and Allinger, 1982) but for small, simple molecules, one's first guess is usually right. Proving that one's MM calculation of a structure and energy is at the global minimum involves proving that no lower energy exists, with all the difficulties attendant to proving a negative. Normally the job is not as difficult as it sounds and for simple molecules, a structure can be shown to be the only plausible one among alternatives. Use of a graphical user interface (e.g., PCMODEL) is recommended for constructing input files as well as for visualizing the output.

In some cases, the steric energy can be used without further calculation. For example, the enthalpy of isomerization of *cis*- to *trans*-2-butene is well known to be about 1.0 kcal mol^{-1} (Chap. 1). A pure bond energy scheme does not distinguish between isomers without some kind of parameter specific to the isomerization written into the program. In MM, however, we recognize that the isomerization enthalpy is nothing more than the difference between steric energies of the isomers, (indirectly dependent of course, on the MM parameterization) which we can easily calculate. The bond energy sums from the MM3 output for the two isomers, are the same, –8.56 kcal mol^{-1}, but when we compare the steric energies from the MM3 output for them, we see that the difference reproduces the enthalpy of isomerization

$$\textit{cis}\text{-2-butene} \rightarrow \textit{trans}\text{-2-butene}$$

$$\Delta_{\text{isom}} H_{298} = \text{steric energy}(\textit{trans}) - \text{steric energy}(\textit{cis})$$

$$= 4.13 - 5.19 = -1.06 \text{ kcal mol}^{-1}.$$

The enthalpy of isomerization as written is negative because the *trans* form is more stable than the *cis*.

If we calculate $\Delta_f H_{298}$ for molecules that are not isomers, contributions from their bond energies (enthalpies) are different. The "normal" bond energies for each molecule are added up and the steric energy is added to the sum to obtain $\Delta_f H_{298}$. A case in point is $\Delta_f H_{298}$ (ethene) for which the MM3 output gives 8.07 kcal mol^{-1} for the summed bond enthalpies, 2.40 as a "partition function contribution" and 2.60 kcal mol^{-1} as the steric energy. The sum is 13.07, which is the MM estimate of the $\Delta_f H_{298}$ (ethene). A similar analysis for ethane but with a bond energy sum of –23.00 kcal mol^{-1} and a different steric energy

(3.39 kcal mol^{-1}) gives $-23.00 + 2.40 + 3.39 = -17.21$ kcal mol^{-1}. The enthalpy of hydrogenation of ethene is

$$\Delta_{hyd}H_{298} = -17.21 - (13.07) = -30.28 \text{ kcal mol}^{-1}.$$

When a -0.42 "torsional correction" to ethane is included we arrive at -30.70 kcal mol^{-1} which compares with Kistiakowsky's experimental value of $\Delta_{hyd}H_{298} = -32.58 \pm 0.05$ kcal mol^{-1}. The discrepancy is well outside the combined error bars of the experiment and the calculation. A calculation by the more recent MM4 force field, however, yields -32.69 kcal mol^{-1} (Allinger *et al.*, 1996). [For more on the partition function contribution and torsional corrections, see Rogers (2003).]

3.5 Molecular Orbital Calculations

Molecular orbital calculations are based on the Schroedinger equation. The full nonrelativistic Schroedinger equation contains $\Psi(\mathbf{r},t)$ where $\mathbf{r}$ is the position vector of a point in Cartesian space and t is the time

$$i\hbar\frac{\partial \Psi(\mathbf{r},t)}{\partial t} = \hat{H}\Psi(\mathbf{r},t). \quad (3.4)$$

By a routine mathematical technique, the equation can be split into a space dependent part and a time dependent part. These two parts of the Schroedinger equation are set equal to the same constant (the energy E) and solved separately. For our purposes, we are interested in the space dependent, time *independent* part, which describes a system that is not in a state of change. In Eq. (3.4), $\hat{H}$ is an operator called the *Hamiltonian operator* by analogy to the classical Hamiltonian function, which is the sum of potential and kinetic energies, and is equal to the total energy for a conservative system

$$H = T + V = E. \quad (3.5)$$

The time independent Schroedinger equation

$$\hat{H}\Psi(\mathbf{r}) = E\Psi(\mathbf{r}) \quad (3.6)$$

has the astonishing property that it applies to *any* mechanical system we might wish to choose because $\hat{H}$ is different for each one. We are interested in the distribution of electrons in space surrounding a collection of atoms which they hold together as a stable molecule.

The principal assumption of the branch of quantum mechanics that we propose to follow is that, for each molecule, there is a wave function $\Psi(\mathbf{r})$ which characterizes all of its physical properties. We are interested in the energy E obtained by operating on $\Psi(\mathbf{r})$ with the Hamiltonian operator $\hat{H}$. Normally, there are many solutions to the time independent Schroedinger equation leading to a *spectrum* of energy levels but we shall be interested in the lowest levels, which, when occupied by electrons, constitute the *ground state* of the molecule. The ground state is the state in which the molecule exists at 298 K. Knowing E for a molecule and its constituent atoms in the standard state, one can calculate $\Delta_f H_{298}$.

The Schroedinger equation cannot be solved exactly except for very simple systems like the hydrogen atom. For molecules, we must be satisfied with an approximate solution of $\hat{H}\Psi(\mathbf{r}) = E\Psi(\mathbf{r})$. In recent years, owing to the work of Pople, Gordon, and others, agreement between MO approximations and such experimental results as exist has been brought to a level that makes quantum thermochemistry competitive with experimental thermochemistry in reliability.

As a first step in obtaining an approximate solution to the molecular Schroedinger equation, we agree to regard the many electron wave function $\Psi(\mathbf{r})$ as having been broken up into molecular orbitals ψ_i

$$\Psi(\mathbf{r}_i) = (n!)^{-1/2} \det[(\psi_1 \alpha)(\psi_1 \beta)(\psi_2 \alpha)...]. \qquad (3.7)$$

The orbitals $\psi_1 \alpha, \psi_1 \beta, \psi_2 \alpha, \ldots$ accommodate single electrons. The symbols α and β designate opposite spins and "det" indicates a *determinant*. Because of spin pairing, the minimum number of molecular orbitals $\psi_1(\mathbf{r}_1), \psi_2(\mathbf{r}_2), \psi_3(\mathbf{r}_3), \ldots$ is one half the number of electrons. The determinant (3.7) is a *Slater determinant*. Computed orbitals beyond the minimum are *virtual orbitals*. They may be sparingly occupied by electrons even in the ground state.

In 1926, Born showed that the probability of finding an electron in an infinitesimal region of space located by its position vector **r** is proportional to the *inner product* of $\Psi(\mathbf{r})$ and its complex conjugate $\Psi^*(\mathbf{r})$

$$\langle \Psi(\mathbf{r}) | \Psi(\mathbf{r}) \rangle. \tag{3.8}$$

[Complex conjugate notation * is redundant and is not used here because the *bra* state vector $\langle \ |$ is already defined as the complex conjugate of the *ket* state vector $| \ \rangle$. The inner product of complex vectors in a *Hilbert space* is a real scalar, as it must be if, as in Born's interpretation, it is to be a probability.]

Because the product (3.8) is a probability, it must integrate to 1.0 when all possible outcomes are taken into account. Consequently, the wave functions are multiplied by arbitrary constants $(n!)^{-1/2}$ chosen to make this integral come out to 1.0 over the complete range of motion. These are called *normalization constants*. It is legitimate to multiply solutions to the Schroedinger equation by an arbitrary constant because they are elements of a *closed binary vector space*. Multiplication of a solution by any scalar yields another element in the space, hence the product of the normalization constant and the wave function (or any other state vector in Hilbert space) also is a solution.

Each time the term "solution" is used in reference to the Schroedinger equation from this point on, the reader should assume that the solution is approximate. According to the *variational principle*, one varies the ψ_i so as to obtain a minimum but approximate E which is an upper bound of the true energy.

In 1951 Roothaan further divided molecular orbitals ψ_i into *linear combinations* of *basis* functions χ_μ

$$\psi_i = \sum_{\mu=1}^{N} c_{\mu i} \chi_\mu \tag{3.9}$$

($\mu = 1, 2, 3, \ldots N$) where the number of one-electron orbitals is greater than the number of electrons, N > n. Having selected a basis set χ_μ, one wishes solve a set of linear simultaneous equations to to find the scalar coefficients $c_{\mu i}$. This mathematical simplification gives a set of *algebraic* equations in place of the set of coupled differential equations in the original problem. Roothaan's equations can be written in matrix form as

$$\mathbf{FC} = \mathbf{SCE} \tag{3.10}$$

where the matrix elements are

$$F_{\mu\nu} = H_{\mu\nu} + \sum_{\lambda\sigma} P_{\lambda\sigma}[(\mu\nu|\lambda\sigma) - (\mu\lambda|\nu\sigma)/2]$$
$$H_{\mu\nu} = \int \chi_\mu H \chi_\nu d\tau$$
$$S_{\mu\nu} = \int \chi_\mu \chi_\nu d\tau$$
$$E_{ij} = \varepsilon_i \delta_{ij}$$

and where

$$P_{\mu\nu} = 2\sum_1^n c_{\mu i} c_{\nu i}$$
$$(\mu\nu|\lambda\sigma) = \iint \chi_\mu(1)\chi_\nu(1) \times \frac{1}{r_{12}} \chi_\lambda(2)\chi_\sigma(2) d\tau_1 d\tau_2$$

with a comparable expression for $(\mu\lambda|\nu\sigma)$. $H_{\mu\nu}$ is the core Hamiltonian that would be imposed by the nuclei on each electron in the absence of all other electrons, and the ε_i are one electron energies. These are the *self consistent field* (SCF) equations.

The integrals $(\mu\nu|\lambda\sigma)$ and $(\mu\lambda|\nu\sigma)$ are difficult to evaluate, which has caused a bifurcation of the field of molecular orbital studies into sub-disciplines followed by those who wish to find suitable empirical constants to substitute for the integrals $(\mu\nu|\lambda\sigma)$ and $(\mu\lambda|\nu\sigma)$ and those who wish to search for better algorithms for solving them. Taking the first of these alternatives, research groups led by Dewar and by Stewart were devoted to obtaining solutions by substituting empirical constants into $F_{\mu\nu}$. The second approach was followed by groups led by Pople, Gordon, and others, who used very efficient computer codes, and relied on the increasing power of contemporary computing machines to solve the integrals in $F_{\mu\nu}$. In general, the rule of speed vs. accuracy has been followed. Semiempirical substitution is faster, hence applicable to larger molecules. *Ab initio* methods are more accurate but they are very expensive in computer resources.

3.6 Semiempirical Methods

Solution of the Schroedinger requires evaluation of many integrals. A large proportion of these integrals make a very small contribution to molecular energy and enthalpy. When they are dropped, the calculation is simplified in the hope that the sacrifice in accuracy will be small. Dropping the integrals in Eq. (3.10) leaves only $H_{\mu\nu}$ in place of the Fock matrix elements $F_{\mu\nu}$. Dropping *some* integrals and replacing others with empirical parameters gives legitimate Hamiltonian elements but elements that are approximate because of the use of empirical parameters. They are elements $H_{\mu\nu}$ in an approximate or semiempirical Hamiltonian matrix. The general rule is that if you are modifying the F matrix to obtain an approximate Hamiltonian, the method is semiempirical. If you are working with the full Hamiltonian and attempting to approach a complete basis set, and thus approach a true wave function, the method is *ab initio*.

3.6.1 *The Huckel method*

Huckel (properly, Hückel) molecular orbital theory is the simplest of the semiempirical methods and it entails the most severe approximations. In Huckel theory, we take the "core" to be frozen so that in the Huckel treatment of ethene, only the two unbound electrons in the p_z orbitals of the carbon atoms are considered. These are the electrons that will collaborate to form a π bond. The three remaining valence electrons on each carbon are already engaged in bonding to the other carbon and to two hydrogens. Most of the molecule, which consists of nuclei, nonvalence electrons on the carbons and electrons participating in the σ

bonds constitute the core. The core serves only to establish an electrostatic field in which the π bonding electrons move.

We now carry out a deceptively simple piece of algebra and make use of the variational principle. The time independent Schroedinger equation (3.6) can be multiplied by $\Psi(\mathbf{r})$ on both sides to give

$$\Psi(\mathbf{r})\hat{H}\Psi(\mathbf{r}) = \Psi(\mathbf{r})E\Psi(\mathbf{r})$$

but E is a scalar, hence we can bring it out of the product on the right and integrate over all space τ

$$\int \Psi(\mathbf{r})\hat{H}\Psi(\mathbf{r})\,d\tau = E \int \Psi(\mathbf{r})\Psi(\mathbf{r})\,d\tau.$$

We divide both sides by the integral of the product $\Psi(\mathbf{r})\,\Psi(\mathbf{r})$, also over all space, to get

$$\frac{\int \Psi(\mathbf{r})\hat{H}\Psi(\mathbf{r})\,d\tau}{\int \Psi(\mathbf{r})\Psi(\mathbf{r})\,d\tau} = E.$$

[Strictly $\langle \Psi(\mathbf{r})|\Psi(\mathbf{r})\rangle$ should be used for the probability density but $\Psi(\mathbf{r})$ is used in place of its complex conjugate because the inner product of both real and complex functions give the same scalar.] E is not the exact energy but, by the variational principle (Mc Quarrie, 1983), E is an upper limit on the energy.

Expansion in a linear combination of atomic orbitals χ leads to a set of integrals that are given the symbols

$$H_{\mu\nu} = \int \chi_\mu(\mathbf{r})\hat{H}\chi_\nu(\mathbf{r})\,d\tau$$

$$S_{\mu\nu} = \int \chi_\mu(\mathbf{r})\chi_\nu(\mathbf{r})\,d\tau.$$

If the basis functions have been selected so that they are normal, their integral over all space is $S_{\mu\nu} = 1.0$ for $\mu = \nu$. In the lowest level approximation, it is common to set $S_{\mu\nu}$ equal to zero for $\mu \neq \nu$, even though we know that the basis functions do not constitute a complete orthonormal set. For ethene the linear combination of the p_z atomic orbitals (which we shall call p_1 and p_2) is

$$\psi = c_1\chi_1 \pm c_2\chi_2 = c_1 p_1 \pm c_2 p_2$$

where c_1 and c_2 are the expansion coefficients. This linear combination of atomic orbitals is called an LCAO approximation.

Minimization in the basis set with respect to the expansion coefficients, which are c_μ and c_ν in the general case, requires that

$$\left(\frac{\partial E}{\partial c_\mu}\right)_\nu = 0 \quad \text{and} \quad \left(\frac{\partial E}{\partial c_\nu}\right)_\mu = 0$$

which leads to a set of simultaneous *algebraic* equations. This set consists of a pair of algebraic equations in the specific case of ethene

$$\left(H_{\mu\mu} - E\right)c_1 + H_{\mu\nu}c_2 = 0 \qquad (3.11a)$$

$$H_{\mu\nu}c_1 + \left(H_{\mu\mu} - E\right)c_2 = 0. \qquad (3.11b)$$

In matrix form,

$$\mathbf{H}_{\mu\nu}\mathbf{C} - E_i\mathbf{IC} = \left(\mathbf{H}_{\mu\nu} - E_i\mathbf{I}\right)\left(\mathbf{C}_\nu\right) = 0 \qquad (3.12)$$

where $\mathbf{I}$ is the diagonal identity matrix, $\mathbf{C}_\nu$ is the vector of expansion coefficients and $E_i\mathbf{I}$ is the diagonal matrix consisting of the spectrum of energy levels E_i spread out along its principal diagonal. There are two vectors of expansion coefficients hence $i = 1, 2$ for ethene. These elements are E_1 and E_2 in the case of ethene.

The *secular* determinantial equation $\left|H_{\mu\nu} - E_i\right| = 0$ because Eqs. (3.11a) and (3.11b) are linearly dependent. We have

$$\begin{vmatrix} \left(H_{\mu\mu} - E_1\right) & H_{\mu\nu} \\ H_{\mu\nu} & \left(H_{\mu\mu} - E_2\right) \end{vmatrix} = 0$$

for ethene. Molecular symmetry is used wherever possible to simplify the mathematics by noticing, for example, that the integral $H_{\mu\nu}$ is the same for both carbon atoms in ethene.

There are two influences on an electron in the p_z orbital of carbon orthogonal to the plane of the molecule. The electron is controlled primarily by the atomic orbital of C and only secondarily by the π molecular orbital in which it participates. In the first case, the matrix element $H_{\mu\nu} = H_{\mu\mu}$ which we define as having an energy α. In the second case $H_{\mu\nu} \neq H_{\mu\mu}$ which we define as having an energy β. In the case of ethene, $H_{\mu\mu} = H_{11} = H_{22} = \alpha$ and $H_{\mu\nu} = H_{12} = H_{21} = \beta$. This makes the orbital secular equation

$$\begin{vmatrix} \alpha - E_1 & \beta \\ \beta & \alpha - E_2 \end{vmatrix} = 0.$$

After dividing each element in the secular determinant by β,

$$\begin{vmatrix} \dfrac{\alpha - E_1}{\beta} & 1 \\ 1 & \dfrac{\alpha - E_2}{\beta} \end{vmatrix} = 0. \qquad (3.13)$$

Expansion of this determinant leads to the binomial equation

$$\left(\frac{\alpha - E_i}{\beta} \right)^2 - 1 = 0$$

with solutions

$$\frac{\alpha - E_i}{\beta} = \pm 1$$

or

$$E_i = \alpha \mp \beta \qquad (i=1,2). \qquad (3.14)$$

The double valued nature of the solution leads to a two level π energy spectrum familiar from elementary treatments.

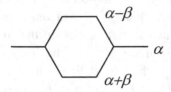

Figure 3.7 Energy diagram of ethene. β is a negative energy.

What remains at this point is evaluation of the parameters α and β. Because Huckel theory is most useful for comparing π energies of similar molecules, the value of α is often assigned a value of 0 or just ignored. It is understood that β is relative to an unspecified value of α, for example, the Huckel eigenvalue spectrum of ethene can be given as $\alpha + \beta, \alpha - \beta$ or simply as $\beta, -\beta$ where β is a definite but unspecified energy. Since β is negative, $\alpha + \beta$ refers to a bonding π orbital with an energy β units below α and $\alpha - \beta$ is an antibonding π^* orbital β units above α. Superscripted * does not indicate a complex conjugate in this context; it is notation for an antibonding orbital. Zero (nonbonding) values of β are also possible in other Huckel molecular orbital calculations.

If desired, the value of β can be estimated spectroscopically by measuring the energy of electromagnetic radiation absorbed when an electron is promoted from the lower level to the upper level in Fig. 3.6. The spectroscopic shift is called a $\pi - \pi^*$ transition. Attempts to relate the spectroscopic energy value (~70 kcal mol^{-1}) to other experimental data are disappointing, as might be expected for this very approximate method. Nor does the Huckel method give a direct estimation of $\Delta_{hyd}H_{298}$(ethene) because we have no specific energy for conversion of 2β for both electrons in the ground state of C–H bonds. With all its quantitative drawbacks, however, Huckel calculations, when extended to larger molecules and to those containing heteroatoms, provided a semiquantitative basis illuminating the research of a generation of organic chemists.

3.6.2 Higher semiempirical methods

A long period elapsed before significant extensions were made to Huckel's 1931 molecular orbital theory. In the 1950s mutual repulsion energies were taken into account by Pariser, Parr, (1953) and Pople (1953) in calculating matrix elements. Hoffmann (1963) devised a scheme whereby the important *off diagonal* matrix elements for σ bonds as well as π bonds were calculated as the arithmetic mean of the adjacent diagonal elements

$$H_{\mu\nu} = 0.88(H_{\mu\mu} + H_{\nu\nu})S_{\mu\nu}$$

where $S_{\mu\nu}$ values were calculated from Slater orbitals and 0.88 is an empirical factor. Slater orbitals are fitted atomic orbitals which retain the exponential form $e^{-\zeta r}$ of Schroedinger's solution for the hydrogen atom, but in which ζ is a variational parameter. The 1960s also saw the rapid spread and almost universal use by quantum chemists of the digital computer.

The most important steps in the further development of computer-based semiempirical methods were in deciding which of the many integrals $(\mu\nu \mid \lambda\sigma)$ and $(\mu\lambda \mid \nu\sigma)$ in the application of Eq. (3.10) to a polyatomic molecule can be dropped, which must be retained and parameterized, and how they should be parameterized. The neglect of differential overlap (NDO) and neglect of diatomic differential overlap (NDDO) approximations, lead to a series of semiempirical programs denoted CNDO (complete neglect of differential overlap), INDO, MINDO, and so on. Initially, NDDO programs were parameterized to reproduce *ab initio* values (see below) for simple molecules but the later semiempirical AM1 programs of (Dewar *et al.*, 1985) and the PM3 method of Stewart (1989) are specifically parameterized against experimental thermochemical results so as to calculate energies and enthalpies. Presently, they are both in wide use for this purpose. Along with thermochemical data, dipole moments, geometries, and isomerization potentials are also calculated by modern semiempirical programs. User's manuals are available on the internet.

In our simple test case for $\Delta_{hyd}H_{298}$(ethene), MNDO calculations yield $\Delta_f H_{298}$ = 15.4 and –19.7 kcal mol^{-1} respectively for ethene and ethane. This gives $\Delta_{hyd}H_{298}$(ethene) = –36.1 kcal mol^{-1} as contrasted to the experimental value of –32.6 ± 0.05 kcal mol^{-1}. PM3 calculations yield $\Delta_f H_{298}$ = 16.6 and –18.1 kcal mol^{-1} for ethene and ethane respectively. This gives $\Delta_{hyd}H_{298}$(ethene) = –34.7 kcal mol^{-1} as contrasted to the experimental value of –32.6 ± 0.05 kcal mol^{-1}.

Thiel (1998) lists mean absolute deviations between calculated $\Delta_f H_{298}$ and experimental $\Delta_f H_{298}$ for two *ab initio* methods, a density functional variant of the *ab inito* method and three semi-empirical methods. They are G2; 1.58, G2(MP2); 2.04, B3LYP; 3.11, MNDO; 9.32, AM1; 7.81, and PM3; 7.01 kcal mol^{-1} respectively. See below for comments on *ab initio* calculations.

Unlike MM, parameterization in semiempirical programs is not specific to each chemical bond type and the problem of incomplete parameterization is not encountered with simple organic compounds. The price for more general parameterization, however, is decreased accuracy. Using the PM3 semiempirical method, the difference between calculated and experimental values of $\Delta_f H_{298}$ (Stewart 1990) is several times as large as it is for MM3. The difference between calculated and experimental results in alkenes and alkynes, our special interest here, becomes larger with increasing unsaturation.

3.7 *Ab initio* Methods

Rigorous and exact solution of the Schroedinger equation would require no empirical parameters at all (beyond mass and charge of the atomic sub particles), and it would be an absolute solution to the problem that can never be changed or revised. In practice, the absolute energy of a molecule is not found, though it may be approached by ever improving approximate methods. Today, what we call *ab initio* procedures retain some degree of empiricism in choice and combination of basis functions, and in small empirical "corrections", but a very serious effort has been put forth to minimize these aspects of the procedure. The limiting factor in rigor is the amount of computer resources one is willing (or able) to expend on a problem, because *ab initio* methods are very computer-intensive.

One condition we impose on the set of functions $\{\chi_\mu\}$ that we choose as a basis set is that they be as nearly complete as possible, that is, we hope that the set will *span* the vector space. In practice, this is never quite achieved. Another condition we would like to build into our basis set is that of easy computation so as to be as economical of computer resources as possible.

3.7.1 *The Gaussian basis set*

About the same time that NDDO approximations were first used, S. F. Boys (1950) showed that the integrals in Eq. (3.10) can be evaluated if the basis set is taken as

$$P(x,y,z)e^{-r^2},$$

that is, as a polynomial $P(x,y,z)$ in Cartesian space times a *Gaussian* function, e^{-r^2}. The result is a *Gaussian type orbital* (GTO) as distinct from the *Slater type orbitals* (STO) previously used. Slater type orbitals are fitted to known wave functions, for example, the $1s$ STO is fitted to the $1s$ wave function of hydrogen, which is exact. The drawback of Slater type orbitals is that they are atomic orbitals, which are similar to but not identical to molecular orbitals. The question is, which *linear combination* can give the best approximation to a molecular orbital, Slater or Gaussian? In the long run, it turned out that the answer is: Gaussian.

It is easy to see that a single Gaussian doesn't represent an s-type atomic orbital as well as the STO because the Gaussian is rounded at $r = 0$ where the atomic orbital has a cusp point. A linear combination of many Gaussians can be made to approach the atomic orbital as closely as we please *near* (but not precisely *at*) the nucleus depending on software efficiency, the power of the computer, and the amount of CPU time we wish to expend. This idea led to programs in which 3 Gaussians approximated a Slater type orbital (the STO-3G approximation), used at the research level until it was supplanted by more accurate programs in the 1980s.

Following the rule that calculated energies can be improved by extending the basis set, one can set up a series STO-2G, STO-3G, STO-4G, and so on, in which an increasing number of Gaussians is used to approximate a Slater type orbital. Unfortunately, we soon find that there is a point of diminishing return, beyond which further elaboration produces little gain. The same is true of Gaussian basis sets arrived at in different ways, for example, 6-31G, except that there are more basis functions and the computation becomes more involved. On the positive side, one finds that the basis functions fall into natural groups, which are replicated from one calculation to the next, hence a way of using computer resources efficiently is by treating each group as though it were a single function.

Each natural group of basis functions, treated as a unit, is called a *contracted Gaussian type orbital* (CGTO). Constituent Gaussians making up a contracted Gaussian are called *primitive Gaussians* or simply *primitives*. If we start out with 36 primitives (by no means a large number in this context) and segment them into groups of 6, we have reduced the computational problem 6-fold.

In *split-valence basis sets*, inner or core atomic orbitals are represented by one basis function and valence atomic orbitals are represented by two. The carbon atom in methane is represented by one $1s$ inner orbital and $2(2s, 2p_x, 2p_y, 2p_z) = 8$ valence orbitals. Each hydrogen atom is represented by 2 valence orbitals, hence the number of orbitals is

$$1 + 8 + 4(2) = 17$$

In the 6-31G basis, the inner shell of carbon is represented by 6 primitives and the 4 valence shell orbitals are represented by 2 contracted orbitals each consisting of 4 primitives, 3 contracted and 1 uncontracted (hence the designation 6-31). That gives us $4(4) = 16$ primitives in the valence shell. The single hydrogen valence shells are represented by 2 orbitals of 2 primitives each. That gives us a total of

$$6 + 4(4) + 4(2 \times 2) = 38 \text{ primitives}$$

which we verify by running the program GAUSSIAN© (gaussian.com) in the 6-31G basis and finding the line

```
17 basis functions    38 primitive gaussians
```

in the output file. The reason the inner shell of carbon is represented by 6 primitives in this basis is that the cusp in the $1s$ orbital is difficult to approximate using Gaussians, which have no cusp.

By the time split basis sets were being used, the effort in *ab initio* molecular orbital theory was no longer to reproduce a Slater type orbital, but to reproduce results of experimental measurements on the chemical and physical properties of the molecule itself. Basis sets were further improved by adding new functions, each representing some refined aspect of the physics of the actual wave function. Chemical bonds are not centered exactly on nuclei, so directionally *polarized* functions were added to the basis set, denoted p, d, or f in such sets as 6-31G(d), 6-31G(d,p), *etc.* Electrons do not have a very high probability density far from the nuclei in a molecule, but the little probability that they do have is important in chemical bonding, so *diffuse functions*, denoted + as in 6-311+G(d) were added in some high level basis sets. While these basis set extensions were being made, the power of computing hardware was growing exponentially to accommodate them.

3.7.2 *Post Hartree–Fock methods*

No matter how good the basis set is made by extension toward an infinite set, one encounters the *Hartree-Fock limit* on the accuracy of molecular energy, because the influence of one electron upon the others has not been fully accounted for in the SCF averaging procedure. The difference between a Hartree-Fock energy and the experimental energy is called the *correlation energy*. To remedy this fault, *correlated* models are made up which consist of a linear combination of the Hartree-Fock solution plus singly, doubly, *etc.* substituted wave functions

$$\psi = a\psi_0 + \sum_{ia} a_i^a \psi_i^a + \sum_{ijab} a_{ij}^{ab} \psi_{ij}^{ab} + \ldots \qquad (3.15)$$

In Eq. (3.15), $ij\ldots$ designate occupied *spinorbitals* (orbitals treated separately according to electron spin α or β) and $ab\ldots$ designate excited orbitals of higher energy than $ij\ldots$, the *virtual orbitals*. Virtual orbitals have small but nonzero occupation numbers. The method is called configuration interaction (CI). The first sum on the right in Eq. (3.15) includes singly substituted orbitals (CIS). Inclusion of the second sum on the right leads to doubly substituted orbitals (CID), while inclusion of both sums is (CISD) and so on. QCISD(T) methods include exponential terms in the expansion and are generally considered to give a better estimate of the energy than the simpler CI terms alone (Pople 1999).

Another method of progressing beyond the Hartree-Fock limit is by inclusion of *many body perturbation terms* (Atkins and Friedman, 1997)

$$E = E_0 + \lambda_i E_i \qquad (i = 1, 2, \ldots) \qquad (3.16)$$

where higher values of $\lambda_i E_i$ lead to correlated energies. Procedures proposed by Møeller and Plesset (1934) are called MP2 and MP4 methods for $i = 2$ or 4 in Eq. (3.16).

3.7.3 *Combined or scripted methods*

Combined or *scripted* programs like the G3(MP2) sequence of Curtiss *et al.* (1999a) and the CBS sequence by Petersson *et al.* (1998) are very popular. In scripted procedures, a short set of instructions (the script) calls up a number of programs to be run in sequence so that their results can be combined in such a way as to produce an energy that is more accurate than the energy from any single calculation. Scripted program suites can be very expensive. A suite of molecular orbital programs developed by Gordon's group and distributed in the nearly forgotten tradition of free exchange of information among practicing scientists is called the *General Atomic and Molecular Electronic Structure System* GAMESS (Schmidt *et al.*, 1993, msg.ameslab.gov).

G3(MP2) calculations are especially well adapted to hydrocarbons, having mean absolute deviations from experiment of < 1.0 kcal mol^{-1} for a large data set. The largest basis set in the G3(MP2) procedure is called, appropriately enough, *G3MP2Large*. The G3(MP2) method uses three calculated points in a basis set-correlation level space to extrapolate to a fourth point, the QCISD(T)/G3MP2Large result, which is inaccessible for the molecule of interest because of limitations on computer time and storage space. The extrapolation is made in two steps. First the MP2/6-31G(d) calculation is "corrected" for basis set inadequacy by carrying out a more rigorous calculation using the MP2/G3MP2Large basis set, taking note of the difference in energy,

$$\Delta E_{MP2} = [E(MP2/G3MP2Large)] - [E(MP2/6-31G(d))]$$

brought about by the calculation at the higher level basis set relative to the lower one. A similar decrement in energy, $[E(QCISD(T)/6-31G(d))] - [E(MP2/6-31G(d))]$, is found upon imposing the post-Hartree-Fock treatment QCISD(T) upon the 6-31G(d) basis set relative to the MP2 energy obtained from the same basis set. Subject to the assumption that the two energy differences are additive, the first correction plus $E(QCISD(T)/6-31G(d))$ gives the desired energy extrapolated to the QCISD(T)/G3(MP2)Large level of approximation.

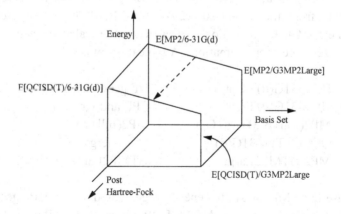

Fig. 3.8 Additive extrapolations in the G3(MP2) scripted method

Details of the new basis set G3MP2Large are given in the original publications (Curtiss et al., 1999a) and it is available on the web (http://chemistry.anl.gov/compmat/g3theory.htm.). Atomic spin-orbit coupling energies $\Delta E(\text{SO})$ are added (C: 0.14 mh, H: 0.0 mh) which, however, cancel in the atomization calculation. A "higher level correction" (HLC) and a zero-point energy ($E(\text{ZPE})$) are added as well. The HLC (9.279 mh = 0.009279 h per pair of valence electrons for a neutral molecule in the ground state) is a purely empirical factor, parameterized so as to give the minimum discrepancy between a large test set of accurately known experimental energies and corresponding calculated energies. The zero point energy arises because the ground state of a quantum harmonic oscillator is one half quantum above the bottom of its parabolic potential well (Mc Quarrie, 1983). The summed zero point energies of all atoms in a molecule, oscillating about their equilibrium positions, is $E(\text{ZPE})$. The sum of these five energy terms is $E_0[\text{G3(MP2)}]$

$$E_0[\text{G3(MP2)}] = E(\text{QCISD(T)}/6-31\text{G(d)}) + \Delta E_{\text{MP2}} \\ + \Delta E(\text{SO}) + E(\text{HLC}) + E(\text{ZPE}) \quad (3.17)$$

The geometry is at the MP2(full)/6-31G(d) level, and the zero-point energy is from the scaled (0.8929) HF/6-31G(d) harmonic oscillator frequencies. Other calculations are carried out on valence electrons only (the "frozen core" approximation). A typical script is

```
HF/6-31G(d) FOpt          HF geometry
HF/6-31G(d) Freq          ZPE and vib freq
MP2(full)/6-31G(d) Opt    MP2(full) geo
QCISD(T)/6-31G(d)         QCI energy
MP2/GTMP2Large            GTMP2Large: energy
```

The base level MP2/6-31G(d) energy is generated at the MP2 geometry in both the QCISD(T) and MP2/GTMP2Large output files. Note well that G3MP2Large is sometimes written G3MP2large or GTMP2Large. These three designations denote the same basis set. The three-point extrapolation energy output files for ethene are

MP2/6-31G(d) = −78.285025
QCISD(T)/6-31G(d) = −78.3221192
MP2/GTMP2Large = −78.3909041

which provide the extrapolated output energy −78.4279983. Upon adding the zero point energy at 0 K in hartrees and correcting by HLC = −6(0.009279) h, we find

$$E_0[\text{G3(MP2)}] = -78.4347736.$$

Upon carrying out the thermal conversion from 0 to 298 K, we obtain the desired enthalpy at 298 K

$$\text{G3MP2 Enthalpy} = -78.430772.$$

The spin-orbit coupling energy has been ignored because of cancellation.

To simplify notation for these two terms let $E_0[\text{G3(MP2)}] \equiv E_0$ and G3MP2 Enthalpy $\equiv H_{298}$. The *thermal correction to the enthalpy* (TCH), converting energy at 0 K to enthalpy at 298, ($H_{298} - E_0 =$ −78.430772 − (−78.4347736) = 0.0040016 h) is a composite of two classical statistical thermodynamic enthalpy changes for translation and rotation, and a quantum harmonic oscillator term for the vibrational energy.

3.7.4 Finding $\Delta_f H_{298}$ and $\Delta_{\text{hyd}} H_{298}$ from G3(MP2) results

There are several ways of converting H_{298} to $\Delta_f H_{298}$ and subsequently to $\Delta_{\text{hyd}} H_{298}$. The "atomization method" is illustrated in Fig. 3.9, which shows a thermochemical cycle for determination of $\Delta_f H_{298}$ of a hydrocarbon. The top horizontal line represents the thermodynamic state of nuclei and electrons, the bottom horizontal line represents elements in their standard states and the verticals, of which there are six, represent enthalpy changes. Each of the three total enthalpy changes in the top half of the figure H_{298} represents a fall from the top state to the state of carbon atoms $H_{298}(C)$, hydrogen atoms $H_{298}(H)$, and the molecule in question $H_{298}(molecule)$, all in the gaseous state. The three verticals at the bottom left of the diagram represent enthalpies of formation $\Delta_f H_{298}$ of C and H from the standard state up to the gaseous state (two steps). The up arrow at the bottom right represents $\Delta_f H_{298}(molecule)$ for the hydrocarbon in question, also in the gaseous state. For the sum of steps about the cycle to be zero, $\Delta_f H_{298}(molecule)$ must be equal and opposite in magnitude to the signed sum of the other 5 steps in the cycle.

$$\Delta_f H^\circ_{298}(molecule) =$$
$$-\left[H^\circ_{298}(C) + H^\circ_{298}(H) - \Delta_f H^\circ_{298}(C) - \Delta_f H^\circ_{298}(H) - H^\circ_{298}(molecule) \right]$$

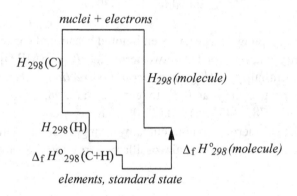

Fig. 3.9 Thermochemical cycle for determination of $\Delta_f H_{298}$ of a hydrocarbon.

In Fig. 3.9, the enthalpy level of the molecule is above that of the elements, that is, $\Delta_f H_{298}$ is positive as in the case of ethene. Ethane, having a negative $\Delta_f H_{298}$ has a molecular energy below that of the elements in a diagram comparable to Fig 3.9. Calculation of $\Delta_f H_{298}$ for ethene and ethane by the atomization method gives 11.9 and -20.1 kcal mol^{-1} leading to $\Delta_{hyd} H_{298}(ethene) = -32.0$ kcal mol^{-1} as contrasted to the experimental value of -32.6 ± 0.05 kcal mol^{-1} at 298 K (Kistiakowsky and Nickle, 1951). The calculated value of $\Delta_f H_{298}(acetylene)$ is 54.3 kcal mol^{-1} which leads to a

$\Delta_{hyd}H_{298}$ (acetylene) of -74.4 kcal mol^{-1} as contrasted to the experimental value of -75.06 ± 0.06 kcal mol^{-1} measured by Kistiakowsky (1938) at 355 K and converted to 298 K by Cox and Pilcher (1970) to give -74.6 ± 0.2 kcal mol^{-1}.

3.7.5 Variation of $\Delta_{hyd}H_{298}$ with T

The definition of the heat capacity at constant pressure

$$C_p = \left(\frac{\partial H}{\partial T}\right)_P$$

leads to the infinitesimal enthalpy change with temperature of a pure substance

$$dH = C_p dT.$$

Over a reasonably short temperature interval ΔT, the equation

$$\Delta H \cong C_p \Delta T$$

is approximately true. The "Thermodynamics" section of the G3(MP2) output file gives $C_V = 7.535$ cal K^{-1} mol^{-1} for ethene and $C_V = 9.488$ cal K^{-1} mol^{-1} for ethane. For an ideal gas,

$$C_P = C_V + R$$

where R is the gas constant, 1.98 cal K^{-1} mol^{-1}, leading to $C_P = 9.52$ cal K^{-1} mol^{-1} for ethene and $C_P = 11.48$ cal K^{-1} mol^{-1} for ethane. These contrast with tabulated experimental values (webbook.nist.gov) $C_V = 10.25$ cal K^{-1} mol^{-1} for ethene and $C_V = 12.55$ J K^{-1} mol^{-1} for ethane. The discrepancies are about 7% in the first case and 8.5% in the second case. This suggests, not surprisingly, that there are modes of motion not accounted for, or that the simple harmonic oscillator approximation of the vibrational part of the G3(MP2) calculation is not quite correct. One also sees that the discrepancy is larger for ethane than it is for ethene, perhaps because of a torsional mode of motion about the carbon-carbon single bond or greater bond angle distortion in the methyl group relative to the methylene group.

Applying Eq. (3.17) to all of the component species of a chemical reaction we get

$$\Delta C_p = \left(\frac{\partial \Delta H}{\partial T}\right)_P \quad (3.18)$$

where

$$\Delta C_P = \sum C_P(\text{products}) - \sum C_P(\text{reactants})$$

$$= 11.48 - 9.52 - 6.90 = -4.94 \text{ cal K}^{-1} \text{ mol}^{-1}$$

as calculated from G3(MP2) for ethene and ethane, taking an experimental value (Atkins, 1994) for the C_P of hydrogen. Calculating ΔC_P from tabulated experimental values for all three constituents of the reaction,

$$\Delta C_P = 12.55 - 10.25 - 6.90 = -4.60 \text{ cal K}^{-1} \text{ mol}^{-1}.$$

The change in $\Delta_r H$ for relatively small changes in T is $\Delta C_P \Delta T$. We can test these two results against Kistiakowsky (1935). Using the computed result for $\Delta C_P \Delta T$ over the temperature range $355-298$, the range between the temperature at which his experiments were carried out (355 K) and standard temperature, we get

$$\Delta \Delta_{\text{hyd}} H_{355} = -4.94\left[-(355-298)\right] = 282 \text{ cal mol}^{-1} = 0.28 \text{ kcal mol}^{-1}$$

from the *computed* value of ΔC_P and $\Delta C_P \Delta T$. Taking the *tabulated* value of ΔC_P, one finds

$$\Delta \Delta_{\text{hyd}} H_{355} = -4.60\left[-(355-298)\right] = 262 \text{ cal mol}^{-1} = 0.26 \text{ kcal mol}^{-1}.$$

These small temperature corrections give

$$\Delta_{hyd}H_{298} = -32.824 + 0.282 = -32.54 \text{ kcal mol}^{-1}$$

and

$$\Delta_{hyd}H_{298} = -32.824 + 0.262 = -32.56 \text{ kcal mol}^{-1}$$

for the enthalpy of hydrogenation of ethene. These two values are to be contrasted with Kistiakowsky's temperature corrected value of $\Delta_{hyd}H_{298} = -32.575 \pm 0.050$ kcal mol^{-1}.

The point of these rather tedious calculations is that the harmonic oscillator approximation brings about a very small error in the thermal correction $\Delta\Delta_{hyd}H_{298}$ in the ethene-ethane hydrogenation. The computed heat capacities are probably reliable over a temperature range of ± 50 K or so. Further, the enthalpy of hydrogen itself is insensitive to temperature so we may take experimental determinations of $\Delta_{hyd}H$ carried out under normal laboratory conditions as essentially the same as $\Delta_{hyd}H_{298}^{\circ}$.

3.7.6 *Isodesmic reactions*

Ab initio enthalpy calculations of $\Delta_f H_{298}$ for hydrocarbons containing more than 2 or 3 carbon atoms are often done by setting up an *isodesmic* reaction in which an experimental $\Delta_f H_{298}$ is known for all participants but one (Hehre, 1970). The enthalpy at 298 K (H_{298}) is calculated for reactants and products, a *computed* $\Delta_f H_{298}$ for the reaction is obtained, and the single remaining unknown $\Delta_f H_{298}$ is calculated from the known experimental values by difference. This scheme was widely used before accurate computed values for E_{298} and H_{298} were available because it is based on cancellation of error across the isodesmic reaction. Computations of $\Delta_f H_{298}$ for individual molecules on the right and left of the equation may suffer considerable error, but under certain conditions, when the algebraic sum of H_{298} values is taken to obtain $\Delta_f H_{298}$ for the reaction, errors cancel because they arise from similar computational defects. The "certain conditions" are evident; one should arrange the chemical reaction in such a way that the reactants have counterpart products as similar in structure and orbital hybridization as possible.

By definition, an isodesmic reaction is one in which the number of bonds and the bond types are the same on both sides of the reaction (Chesnut, 1996). An illustration of the use of isodesmic reactions is from the work of Cheng and Li (2003) in which they determined $\Delta_f H_{298}$ of n-*tert*-butyl methanes, where n = 1, 2, 3, and 4. We shall take the simplest example, calculation of $\Delta_f H_{298}$ of *tert*-butylmethane (2,2-dimethylpropane) itself. The isodesmic reaction is

$$\begin{array}{c} CH_3 \\ | \\ CH_3\text{-}C\text{-}CH_3 \\ | \\ CH_3 \end{array} + 3CH_4 \longrightarrow 4C_2H_6$$

$\quad$ −197.35354 $\quad\quad$ −40.42210 $\quad\quad$ −79.65120
$\quad\quad\quad\quad\quad\quad\quad\quad$ −17.8 $\quad\quad\quad\quad$ −20.1

where computed values of H_{298} are in the first line below the reaction, and experimental values of $\Delta_f H_{298}$ for CH_4 and C_2H_6 are in the second line. On the reactant side, two alkanes have a total of 24 C−H bonds and 4 C−C bonds. On the product side, one molecule has a total of 4(6) = 24 C−H bonds and 4(1) = 4 C−C bonds. One value is missing, from this scheme, that of $\Delta_f H_{298}$ for *t*-butylmethane.

The *calculated* enthalpy of reaction $\Delta_r H_{298}$ is

$$\Delta_r H_{298} = 4(-79.64672) - (-197.35354) - 3(-40.41828)$$

$$\Delta_r H_{298} = 0.02150 \text{ hartrees} = 13.49 \text{ kcal mol}^{-1}.$$

This enthalpy change is used to calculate the remaining unknown from the *experimental* values of $\Delta_f H_{298}$ for methane and ethane

$$13.5 = 4(-20.1) - \Delta_f H_{298}(t\text{-butylmethane}) - 3(-17.9)$$

which leads to

$$\Delta_f H_{298}(t-\text{butylmethane}) = -40.2 \text{ kcal mol}^{-1}$$

as contrasted to the experimental value of -39.9 ± 0.2 kcal mol^{-1}.

The authors carry these calculations on in the logical sequence of di(t-butyl)methane (-59.2), tri(t-butyl)methane (-55.3), and tetra(t-butyl)methane, (-33.7 kcal mol^{-1}). Agreement with the experimental value is good for the second compound in the series, but experimental work is uncertain and under debate for the third compound. An experimental value is nonexistent for the last compound named, which is highly strained, and has not yet been synthesized. Uncertainty in the experimental work on the third compound in this series and lack of experimental data on the fourth is a result of the extreme steric repulsion suffered by three or four t-butyl groups crowded about the same carbon atom. The minimum in $\Delta_f H_{298}$ for these four compounds (−40.2, −59.2, −55.3, −33.7 kcal mol^{-1}) arises as strain energy contributes a larger component of the total $\Delta_f H_{298}$ for the latter compounds in the series. Addition of a second t-butyl group to (t-butyl)methane brings about a decrease in $\Delta_f H_{298}$, and establishes a negative trend that would be continued for addition of a third and fourth t-butyl group but for the steric strain which is so severe as to reverse the trend and cause a minimum at about tri(t-butyl)methane, followed by an increase in $\Delta_f H_{298}$ between tri- and tetra(t-butyl)methane. This is a good example of verification of a method for known compounds followed by extension to unknowns that are not amenable to experimental work.

If H_{298} is computed for a number of molecules related to one another by hydrogenation and isomerization reactions as in Fig. 3.10, and if $\Delta_f H_{298}$ is known for any one of them, $\Delta_f H_{298}$ can be calculated for all the rest (Rogers et al., 1998). In Fig. 3.10, the experimental value of the enthalpy of formation of acetone is well known, $\Delta_f H_{298} = 52.23 \pm 0.14$ (Wiberg et al., 1991), as are one or two of the other values, but most are not. Calculation of $\Delta_f H_{298}$ for the remaining members in the scheme depends upon reactions with similar reactants and products (though not necessarily isodesmic) plus computed enthalpies of isomerization and hydrogenation.

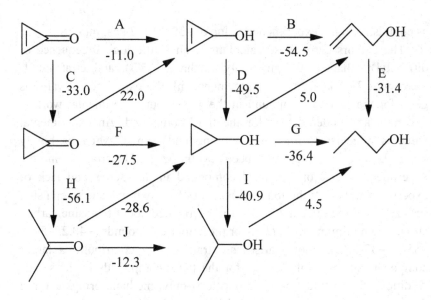

Fig. 3.10 A reaction scheme for related ketones and alcohols. Acetone is at the bottom left. $\Delta_f H_{298}$ for any member of this scheme can be calculated from $\Delta_f H_{298}$ (acetone) = 52.2 kcal mol^{-1} by summing the $\Delta_r H_{298}$ over an appropriate reaction path within the scheme.

Like isodesmic reactions, hydrogenations involve similar species as reactant and product, the only change upon reaction being saturation of one or more multiple bonds. Unlike isodesmic reactions, however, hydrogenations involve a change in orbital hybridization, which may interfere with error cancellation. An advantage that a hydrogenation-isomerization scheme has over isodesmic reactions is that isodesmic reactions may have large stoichiometric coefficients, for example, the isodesmic reaction for determination of tetra(*t*-butyl) methane is

$$C(t\text{-butyl})_4 + 15\ CH_4 \rightarrow 16\ C_2H_6$$

Whatever the error there may be in an experimental measurement of $\Delta_f H_{298}$ of the smaller molecules in an isodesmic reaction, it is multiplied by their stoichiometric coefficients in calculating $\Delta_f H_{298}$ of the target molecule. For example, whatever small error there may be in $\Delta_f H_{298}$ of ethane in the example cited above, it is multiplied by 16 in the calculation.

Hydrogenation reactions have a stoichiometric coefficient of 1 for the reactant, 1 for the product, and 1 for H_2 in hydrogenation of monoenes, 2 for dienes or alkynes, and so on, to a relatively small number, even in the rare examples of hydrogenation of polyalkenes (Rogers et al., 2005). Isomerizations are even simpler, having 1:1 stoichiometric coefficients. Thus, in some cases, the advantage of small stoichiometric coefficients may make a hydrogenation-isomerization scheme the method of choice in obtaining $\Delta_f H_{298}$ values from computed H_{298}.

3.7.7 Ionization potentials, proton affinities and electron affinities

Ionization potentials can be found from the difference between the energy of a neutral species and that of its positive ion. In the critical case of the H atom, the ionization potential is equal to the energy of the electron in the ground state (but with the opposite sign). Schroedinger showed us how to calculate this *exact* energy from fundamental constants as 13.6057 electron volts eV which is 0.5000 h ≡ 313.7545 kcal mol^{-1} by the definition of the atomic unit of energy h as one hartree. A G3(MP2) calculation of the ground state electronic energy of the hydrogen atom yields 314.9 kcal mol^{-1}.

Calculation of the ground state energy of the helium atom is a critical case as well because it is the first example of correlation energy, the difference between the Hartree-Fock energy and the exact value. The energy required to remove one electron from a neutral He atom is the *first ionization potential*

$$He \rightarrow He^+ + e^-$$

so called because it is the potential energy of attraction of He^+ for an electron that must be overcome to free the electron e^-.

The exact value cannot be calculated because of the insolubility of the three-body problem of two electrons and a doubly charged nucleus. Nevertheless, we do know the exact energy necessary to remove the second electron

$$He^+ \rightarrow He^{2+} + e^-$$

because it is equal and opposite to the binding energy of a single electron in a field of $Z = 2$ atomic units, that is, $2^2 = 4$ times the electronic energy of the hydrogen atom or 2.0000 h. The calculated value of the total electronic energy of the He atom minus 2.0000 is its computed first ionization potential.

The ionization energy of both electrons in He can be measured experimentally. The value of 2.9033 h for total ionization is thought to be very accurate. The total ionization energy minus 2.0000 is the experimental value of the binding energy of He^+ for an electron: –0.9033 h (with a change in sign).

The total *approximate* ionization energy of He can also be calculated. The result of a Hartree-Fock calculation is –2.8617 for the total energy of He, hence –0.8617 h is the calculated value for the one-electron binding energy. The difference between the Hartree-Fock calculation and the experimental value is the correlation energy, –0.9033 – (–0.8617) = –0.0416 h = –26.10 kcal mol^{-1}. Clearly a discrepancy of this size is unacceptable for any chemical application, where a typical bond energy is only about 100 kcal mol^{-1}. The object of post Hartree-Fock calculations, such as QCISD(T), MP2 *etc.* is reduction or removal of this correlation energy from the final result. In computing ionization potentials IPs, a practical target accuracy is ± 2 kcal mol^{-1} or better. The scripted G3(MP2) calculation is largely successful in achieving this goal for H and He, yielding an ionization potential for H of 314.9 kcal mol^{-1} (discrepancy –1.1 kcal mol^{-1}) and 566.2 = 0.9023 h (discrepancy 0.0010 h = 0.6 kcal mol^{-1}) for the first ionization potential of He.

The first ionization potential of CH_4

$$CH_4 \rightarrow CH_4^+ + e^-$$

is 291.8 kcal mol (discrepancy 0.8 kcal mol^{-1}) and for ethene, it is 244.5 (discrepancy. −2.2 kcal mol^{-1}).

The *proton affinity* is the energy change for the reaction

$$A + H^+ = AH^+$$

taking $\Delta_f H_{298}(H^+) = 0$, that is, the proton affinity is the energy difference between $\Delta_f H_{298}$ for (AH$^+$) and $\Delta_f H_{298}$(A) for an arbitrary proton attractor A. Similarly, the electron affinity is the energy change

$$A + e^- \rightarrow A^-$$

for addition of an electron to a molecular or atomic species.

3.7.8 Density functional theory (DFT)

Density functional theory is based on the fact that the average energy of an electron in an atom or a molecule is a function of its probability density ρ in the vicinity of the nucleus or nuclei, and ρ is a function of its position in space. A function of a function is a *functional*. In DFT, the energy of an electron is expressed as a sum of three classical energies, a kinetic energy T, a nuclear charge repulsion potential V_n, and an electron charge repulsion potential V_e, plus one quantum mechanical energy E^{XC}

$$E = T + V_n + V_e + E^{XC}. \qquad (3.19)$$

The *Kohn-Sham* equations are very like the Schroedinger equations

$$\widehat{K}\psi_i = E_i \psi_i$$

(Use of the plural in referring to these equations indicates a *set* of equations leading to a *spectrum* of energy levels.) The Kohn-Sham operator $\widehat{K}$ contains three classical operators corresponding to the energies in Eq. (3.19), plus E^{XC}. For an electron, designated electron 1,

$$K_1 = -\frac{1}{2}\nabla_1^2 - \sum_a \frac{Z_a}{r_{1a}} + \sum_j \int \rho_j \frac{1}{r_{1j}} d\tau_j + E^{XC}$$

The classical terms are readily computed but difficulty is encountered in calculating E^{XC} called the *exchange-correlation* energy.

The exchange-correlation functional can be divided into two parts, the *exchange energy* E^X, and the *correlation energy* E^C

$$E^{XC} = E^X + E^C.$$

Several schemes exist for calculating E^{XC}. All are largely empirical, containing adjustable parameters, though each rests on a reasonable theoretical model. One of the most accurate functionals for hydrocarbon thermochemistry, as indicated by comparison of calculated $\Delta_f H_{298}$ values with experiment, was devised by Becke (1988) and Lee, Yang, and Parr (1988), hence the combination is called a BLYP theory. The B3LYP method uses a hybrid functional containing 3 parameters.

DFT methods, specifically the B3LYP protocol, are quite fast relative to other accurate molecular orbital model chemistries because of the ease in calculating the classical terms in the Kohn-Sham operator. For this reason they have enjoyed popularity among computational chemists working on practical problems. The drawback of DFT methods is that, as yet, no systematic approach to an exact wave function and energy is known. This contrasts with G3 theories which, in principle, approach exact solutions as larger basis sets and more sophisticated correlation methods are devised. Hence we can expect a systematic improvement from Gaussian theories but DFT methods may or may not improve with future theoretical work. At present, B3LYP calculations are quite satisfactory for smaller hydrocarbons but do not compete as successfully in studies of those larger hydrocarbons for which experimental data permit comparison.

Density functional methods introduce more empiricism into the calculation than G3(MP2) or CBS (below) and are less legitimately described as *ab initio* than Gaussian methods because of the arbitrary parameters included in estimating E^{XC}. Therefore they occupy a position somewhere between the strictly *ab initio* methods and the semi-empirical protocols. Even in the most parameterized versions, however, the number of parameters is far smaller than that of the empirical or semi-empirical methods such as AM1 and PM3.

DFT methods have been used by Pan *et al.* (1999) to study energies of diverse hydrocarbons. They found $\Delta_f H_{298}$ for ethene and ethane 12.0 and -20.1 kcal mol^{-1} $\Delta_{hyd} H_{298}$ of -32.1 kcal mol^{-1} as contrasted to the experimental value of -32.6 kcal mol^{-1}.

3.7.9 *Complete basis set extrapolations (CBS)*

Complete basis set extrapolation programs are, like the G3 programs, scripted. The exceptional step is the basis set extrapolation itself, in which it is noticed that the contribution to chemical energies by higher order perturbation terms decreases relative to lower order terms. Thus computer resources can be used most effectively by carrying out a series of calculations in which high order basis sets (computer intensive) are paired with low order perturbational calculations (not computer intensive) and *vice versa*. A very high level calculation is carried out at a low perturbation order, a second calculation is carried out in which the basis set expansion is less complete but the perturbation order is higher, and so on until the highest order perturbation term is obtained at a very low basis set expansion. In the example of a CBS-q calculation, the basis sets are 6-311+G(3d2f,2df,2p) at the SCF level, 6-31+G″ at MP2, 6-31G′ at MP4(SDQ), and 6-31G at QCISD(T). (Basis sets designated 6-31+G″ and 6-31G′ are modifications of the corresponding Gaussian G* basis sets.)

Components are selected for each model chemistry so that no step dominates the others in either computer time consumed or error suffered in the final result.. A recent member of the CBS family, CBS-QB3, uses the CBS-Q script carried out at the B3LYP geometry using the B3LYP ZPE. CBS-QB3 appears to be quite competitive with G3(MP2) calculations in speed and accuracy. CBS-QB3 calculations of $\Delta_f H_{298}$ for ethene and ethane yield 7.17 and -23.24 and kcal mol^{-1} respectively.

These result in a calculated $\Delta_{hyd}H_{298}$ of −30.4 kcal mol⁻¹ as contrasted to the experimental value at 298 K of −32.6 kcal mol⁻¹.

3.7.10 Bond dissociation energies

Free radicals are important in atmospheric systems, flames, explosions, polymerizations, and biological systems. They have one or more free electrons that result from *homolytic* cleavage of normal chemical bonds. Free radicals are very active and typically appear as intermediates on the reaction path from one stable species to another, hence their experimental study is difficult. Like $\Delta_f H_{298}$ of positive and negative ions, $\Delta_f H_{298}$ of free radicals can be computed *ab initio*.

The bond dissociation enthalpy (*BDE*) for a diatomic molecule AB is the amount of energy necessary to bring about homolytic cleavage, that is, the enthalpy of the dissociation reaction

$$AB \rightarrow A\cdot + B\cdot.$$

In the case of ethane, homolytic cleavage of the C–C bond

$$CH_3CH_3 \rightarrow 2CH_3\cdot$$

has a $\Delta_{dissoc}H_{298}$ and a BDE

$$BDE[CH_3CH_3] = \Delta_{dissoc}H_{298}[CH_3CH_3]$$
$$= 2\Delta_f H_{298}[CH_3\cdot] - \Delta_f H_{298}[CH_3CH_3].$$

Model chemistries can be combined. One of many combined model chemistries is the G3(MP2)B3 model which consists of a G3(MP2) calculation carried out at the B3LYP geometry and which uses the B3LYP zero point energy. Combination is achieved by replacing the first three members of the G3(MP2) script with the corresponding B3LYP optimization steps, leaving the rest of the script unchanged. The *BDE* for ethane in the G3(MP2)B3 model chemistry is

$$BDE[CH_3CH_3] = 2(-39.754860) - (-79.651012) = 0.141292 \text{ h}.$$

The *BDE* of ethane to produce methyl radicals is a simple example of the more general case for hydrocarbons and their radicals

$$CH_3R \rightarrow CH_3 \cdot + R \cdot$$

$$BDE[CH_3R] = \Delta_f H_{298}[CH_3 \cdot] + \Delta_f H_{298}[R \cdot] - \Delta_f H_{298}[CH_3R].$$

In the case of the propargyl radical $\cdot CH_2C \equiv CH$, the *BDE*, also in the G3(MP2)B3 model chemistry, is

$$BDE[HC \equiv CCH_2 - CH_3] = 0.124404 \text{ h}.$$

One would like to have a quantitative measure of the stability (or instability!) of these very active species. The traditional method for ordering alkane radicals rests on the *BDE*s of the C – H bonds, which are in the order

$$BDE[CH_3 - H] > BDE[CH_3CH_2 - H]$$

$$> BDE[(CH_3)_2 CH_2 - H] > BDE[(CH_3)_3 CH_2 - H].$$

This leads to the correct sequence of stabilities among the radicals: tertiary > secondary > primary > methyl on the premise that the smaller the *BDE*, the more stable is the product alkane radical formed in the endothermic C–H bond breaking reaction. Though the sequence is correct for the alkanes listed, it is logically and quantitatively incorrect according to Zavitsas (2001) and Coote *et al.* (2003). They argue that the traditional calculation is logically incomplete in that it ignores differences in polarity of the C–H bond in different types of alkanes. Different ordering of alkyl radical stabilities is obtained if one tries to apply the same reasoning and calculation to a corresponding sequence of alcohols, alkyl halides, *etc*.

Calculations based on the *BDE* of the C–C bond in alkanes do not suffer this disadvantage because the polarity effect is minimized or nonexistent. The solution to the problem is to calculate *stabilization energies SE* of the general radical R· *relative to the methyl radical*

$$SE[R\cdot] = BDE[CH_3CH_3] - BDE[CH_3R].$$

The difference between the two *BDE* equations for breaking C–C bonds is the *SE* of radical R· relative to that of $CH_3\cdot$

$$SE[R\cdot] = 2\Delta_f H_{298}[CH_3\cdot] - \Delta_f H_{298}[CH_3CH_3]$$
$$-\{\Delta_f H_{298}[CH_3\cdot] + \Delta_f H_{298}[R\cdot] - \Delta_f H_{298}[CH_3R]\}.$$

Cancellation of two $\Delta_f H_{298}[CH_3\cdot]$ terms results in some simplification

$$SE[R\cdot] = \Delta_f H_{298}[CH_3\cdot] - \Delta_f H_{298}[CH_3CH_3]$$
$$-\Delta_f H_{298}[R\cdot] + \Delta_f H_{298}[CH_3R].$$

The first two terms in the sum on the right are the same for any alkyl radical relative to $CH_3\cdot$ so we can calculate it once and for all and label it *K*. Now

$$SE[R\cdot] = K - \Delta_f H_{298}[R\cdot] + \Delta_f H_{298}[CH_3R]$$

where

$$K = \Delta_f H_{298}[CH_3\cdot] - \Delta_f H_{298}[CH_3CH_3]$$
$$= -39.754860 - (-79.651012) = 39.896152 \text{ h}$$

in the G3(MP2)B3 model chemistry.

Continuing with the example of propargyl relative to the methyl radical,

$$SE[\cdot CH_2C \equiv CH] = 39.896152 - \Delta_f H_{298}[\cdot CH_2C \equiv CH]$$

$$+\Delta_f H_{298}[HC \equiv CCH_2CH_3]$$

$$= 0.016888 \text{ h} = 10.60 \text{ kcal mol}^{-1}.$$

The propargyl radical is 10.6 kcal mol^{-1} more stable than the methyl radical in the G3(MP2)B3 model chemistry, presumably because of increased electron delocalization over the whole $\cdot CH_2C \equiv CH$ radical as contrasted to restricted localization of the free electron in $CH_3 \cdot$. Analogous calculations in the G3(MP2), CBS-QB3 etc. model chemistries give answers that are quite close to the one just shown for the G3(MP2)B3 model chemistry, as they must if we are to believe that calculation of $\Delta_f H_{298}$ of alkanes and alkyl radicals gives a good approximation to the true value.

Calculations are best carried out in units of hartrees, then converted to kcal mol^{-1} if desired (conversion constant 627.51 kcal mol^{-1}/hartree) to avoid needless confusion over slightly differing literature values of C and H atomization energies and of the corresponding enthalpies of formation of the free atoms from the elements in their standard states. One can do this because the C and H atomization energies and enthalpies of formation cancel in the full calculation.

Bibliography

Alberty, R. A.; Silbey, R. J. (1996). *Physical Chemistry*, 2nd. ed. Wiley, New York.
Allinger, N. L. (1976). Calculation of molecular structure and energy by force-field methods, *Advances in Phys. Org. Chem.*, 13, pp. 1–82.
Allinger, N. L.; Dodziuk, H.; Rogers, D. W.; Naik, S. N. (1982). Heats of hydrogenation and formation of some 5-membered ring compounds by molecular mechanics calculations and direct measurement, *Tetrahedron*, 38, pp. 1593–1597.
Allinger, N. L.; Yuh, Y. H.; Lii, J.-H. (1989). Molecular mechanics. The MM3 force field for hydrocarbons. 1, *J. Am. Chem. Soc.*, 111, pp. 8551–8566.
Allinger, N. L.; Lii, J.-H. (1989). Molecular mechanics. The MM3 force field for hydrocarbons. 2. Vibrational frequencies and thermodynamics, *J. Am. Chem. Soc.*, 111, pp. 8566–8575.
Allinger, N. L.; Chen, K.; Lii, J.-H. (1996). An improved force field (MM4) for saturated hydrocarbons, *J. Comp. Chem.*, 17, pp. 642–668.
Anderson, J. M. (1966). *Mathematics for Quantum Chemistry*, Benjamin, New York.
Atkins, P. W. (1998). *Physical Chemistry*, 6th ed. Freeman, New York.
Atkins, P. W.; Friedman, R. S. (1997). *Molecular Quantum Mechanics*, 3rd ed. Oxford Univ. Press, Oxford.
Baerends, E. J.; Gritsenko, O. V. (1997). A quantum view of density functional theory, *J. Phys. Chem.*, 101, pp. 5383–5403.
Barrante, J. R. (1998). *Applied Mathematics for Physical Chemistry*, 2nd ed. Prentice-Hall, Engllewood Cliffs, NJ.
Barrow, G. M. (1996). *Physical Chemistry*, 6th ed. WCB/Mc Graw-Hlll, New York.
Becke, A. D. (1988). Density-functional exchange energy approximation with correct asymptotic behavior, *Phys. Rev A*, 38, pp. 3098–3100.
Benson, S. W. (1976). *Thermochemical Kinetics*, 2nd ed., Wiley, New York.
Benson, S. W.; Cohen, N. (1998). Current Status of Group Additivity. In Irikura, K. K.; Frurip, D. J. eds., *Computational Thermochemistry*, ACS Symposium Series 677, American Chemical Society, Washington D. C.
Born, M. (1926). Zur Quantenmechanik der Stoßvorgänge, *Zeit. f. Physik*, 37, pp. 863–867.
Born, M. (1926). Quantenmechanik der Stoßvorgänge, *Zeit. f. Physik*, 38, pp. 803–827.
Boys, S. F. (1950). Electronic Wave Functions. I. A general method of calculation for the stationary states of any molecular system, *Proc. Royal Soc. London.*, Ser. A., 200, pp. 542–554.

Bretschneider, E.; Rogers, D. W. (1970). *Mikrochim. Acta [Wien]* pp. 482–490.
Burkert, U.; Allinger, N. L. (1982). *Molecular Mechanics*, ACS Publ. No. 177, American Chemical Society. Washington D. C.
Caldwell, R. A.; Liebman, J. F.; Rogers, D. W.; Unett, D. J. (1997). Enthalpies of hydrogenation and of formation of 1-phenyl cycloalkanes, *J. Molec. Structure*, 413–414 pp. 575–578.
Carnahan, B.; Luther, H. A.; Wilkes, J. O. (1969). *Applied Numerical Methods*, Wiley, New York.
Chase, M. W. Jr.; Davies, C. A.; Downey, J. R. Jr.; Frurip, D. J.; McDonald, R. A.; Syverud, A. N. (1985). *J. Phys. Chem. Ref. Data*, 14, Supp 1.
Chase, M. W. Jr., (1998). *NIST-JANAF Themochemical Tables*, 4th ed., *J. Phys. Chem. Ref. Data, Monograph* 9, 1.
Chatterjee, S.; Price, B. (1977). *Regression Analysis by Example*, Wiley, New York.
Cheng M.-F.; Li, W.-K. (2003). Structural and energetics studies of tri- and tetra-*tert*-butylmethane, *J. Phys. Chem.*, 107, pp. 5492–5498.
Chesnut, D. B.; Davis, K. M. (1996). Resonance revisited: A consideration of the calculation of cyclic conjugation energies, *J. Comput. Chem.*, 18, pp. 584–593.
Christensen, J. J.; Gardner, J. W.; Eatough, D. J.; Izatt, R. M.; Watts, P. J.; Hart, R. M. (1973). An Isothermal Titration Microcalorimeter, *Review of Scientific Instruments*, 44, pp. 481–484.
Cohen, N.; Benson, S. W. (1993). Estimation of heats of formation of organic compounds by additivity methods, *Chem Rev.*, 93, pp. 2419–2438.
Conant, J. B.; Kistiakowsky, G. B. (1937). Energy changes involved in the addition reactions of unsaturated hydrocarbons, *Chem. Rev.*, 20, pp. 181–194.
Conn, J. B.; Kistiakowsky, G. B.; Smith, E. A. (1939). Heats of organic reactions. VIII. Some further hydrogenations, including those of some acetylenes, *J. Am. Chem. Soc.*, 61, pp. 1868–1876.
Coote, M. L.; Pross, A.; Radom, L. (2003). Variable trends in R − X bond dissociation energies (R=Me, Et, *i*-Pr, *t*-Bu), *Organic Letters*, 5, pp. 4689–4692.
Cox, J. D. and Pilcher, G. (1970). *Thermochemistry of Organic and Organometallic Compounds*, Academic Press, London.
CRC Handbook of Chemistry and Physics, (1993). Lide, D. R., ed., CRC Press, Boca Raton, FL.
Curtiss, L. A.; Raghavachari, K.; Trucks, G. W.; Pople, J. A. (1991). Gaussian-2 theory for molecular energies of first- and second-row compounds, *J. Chem. Phys.*, 94, pp. 7221–7230.
Curtiss, L. A.; Redfern, P. C.; Smith, B. J.; Radom, L. (1996). Gaussian-2 theory: reduced basis set requirements, *J. Chem. Phys.*, 104, pp. 5148–5152.
Curtiss, L. A.; Raghavachari, K.; Redfern, P. C.; Pople, J. A. (1997). Assessment of Gaussian-2 and density functional theories for the computation of enthalpies of formation, *J. Chem. Phys.*, 106, pp. 1063–1079.

Curtiss, L. A.; Redfern, P. C.; Raghavachari, K.; Rassolov, V.; Pople, J. A. (1999a). Gaussian-3 theory using reduced Møller-Plesset orders, *J. Chem. Phys.*, 110, pp. 4703–4709.

Curtiss, L. A.; Redfern, P. C.; Raghavachari, K.; Baboul, A. G.; Pople, J. A. (1999b). Gaussian-3 theory using coupled cluster energies, *Chemical Physics Letters*, 314, pp. 101–107.

Dence, J. B. (1975). *Mathematical Techniques in Chemistry*, Wiley, New York.

Dewar, M. J. S. (1969). *The Molecular Orbital Theory of Organic Chemistry*, Mc Graw-Hill, New York.

Dewar, M. J. S. (1975). Quantum organic chemistry, *Science*, 187, pp. 1037–1044.

Dewar, M. J. S.; Storch, D. (1985a). Development and use of quantum molecular models. 75. Comparative tests of theoretical procedures for studying chemical reactions, *J. Am. Chem. Soc.*, 107, pp. 3898–3902.

Dewar, M. J. S.; Zoebish, E.; Healy, E. F.; Stewart, J. J.P. (1985b). AM1: A new general purpose quantum mechanical molecular model, *J. Am. Chem. Soc.*, 107, pp. 3902–3909.

Dewar, M. J. S.; Thiel, W. (1977). Ground states of molecules. 38. The MNDO method. Approximations and parameters, *J. Am. Chem. Soc.*, 99, pp. 4899–4907.

Dickson, T. R. (1968). *The Computer and Chemistry*, Freeman, San Francisco.

Dirac, P. A. M. (1929). *Proceedings Royal Soc. London*, A, 123, pp. 714–733.

Doering, W. von E.; Roth, W. R.; Breuckman, R.; Figge, L.; Lennartz, H.-W.; Fessner, W.-D.; Prinzbach. H. (1988). Zur Frage der Homoaromatizität von Norcaradien und Cycloheptatrien, *Chem. Ber.*, 121, pp. 1–9.

Doering, W. von E.; Roth, W. R.; Bauer, F.; Boenke, M.; Breuckmann, R.; Ebbrecht, M; Herbold, M.; Schmidt, R.; Lennartz, H-W.; Lenoir, D.; Boese, R. (1989). Rotationsbarrieren gespannter Olefine, *Chem. Ber.*, 122, pp. 1263–1275.

Dolliver, M. A.; Gresham, T. L.; Kistiakowsky, G. B.; Vaughan, W. E. (1937). Heats of organic reactions. V. Heats of hydrogenation of various hydrocarbons, *J. Am. Chem. Soc.*, 59, pp. 831–841.

Dolliver, M. A.; Gresham, T. L.; Kistiakowsky, G. B.; Smith, E. A.; Vaughan, W. E. (1938). Heats of organic reactions. VI. Heats of hydrogenation of some oxygen-containing compounds, *J. Am. Chem. Soc.*, 60, pp. 440–450.

Ebbing, D. D.; Gammon, S. D. (1999). *General Chemistry*, Houghton Mifflin, Boston.

Ebert, K.; Ederer, H.; Isenhour, T. L. (1989). *Computer Applications in Chemistry*, VCH Publishers, New York.

Eğe, S. N. (1994). *Organic Chemistry; Structure and Reactivity*, 3rd ed. D. C. Heath, Lexington, Mass.

Fang, W.; Rogers, D. W. (1992). Enthalpy of hydrogenation of the hexadienes and *cis*- and *trans*-1,3,5-hexatriene, *J. Org. Chem.*, 57, pp. 2294–2297.

Fishtik, I.; Datta, R. (2003). Aromaticity vs. stoichiometry, *J. Phys Chem. A*, 107, pp. 10471–10476.

Flury, P.; Grob, C. A.; Wang, G. Y.; Lennartz, H-W.; Roth, W. R. (1988). Bridging Strain in Norbornyl and Oxanorbornyl Cations, *Helvitica Chimica Acta*, 71, pp. 1017–1024.

Fock, V. A. (1930). Naherungsmethode zur Lösung des quantenmechanischen Mehrkorperproblems, *Physik.*, 61, pp. 126–148.

Foresman, J. B.; Frisch A. (1996). *Exploring Chemistry with Electronic Structure Methods*, 2nd ed. Gaussian Inc. Pittsburgh.

Glukhovtsev, M. N.; Laiter, S. (1995a). High level *ab initio* stabilization energies of benzene, *Theor. Chim. Acta*, 92, pp. 327–332.

Glukhovtsev, M. N.; Laiter, S.; Pross, A. (1995b). Thermochemistry of Cyclobutadiene and Tetrahedrane: A High-Level Computational Study, *J. Phys. Chem.*, 99, pp. 6828–6831.

Grant, G. H.; Richards, W. G. (1995). *Computational Chemistry*, Oxford Sci. Pub. New York.

Greenwood, H. H. (1972). *Computing Methods in Quantum Organic Chemistry*, Wiley Interscience, New York.

Grimme, W.; Grommes, T.; Roth, W. R.; Breuckmann, R. (1992). The Benzene Ring as Dienophile in an Intramolecular [4+2] Cycloaddition: Degenerate Rearrangement of 7,8-Benzobicyclo[4.2.2]deca-2,4,7,9-tetraene, *Angew. Chem. Int. Ed. Engl.*, 31, pp. 872–874.

Halgren, T. A.; Nachbar, R. B. (1996). *Merck Molecular Force Field I - V*, J. Computational Chemistry, 17, pp. 490–615.

Hanna, M. W. (1981). *Quantum Mechanics in Chemistry*, 3rd ed. Benjamin, Menlo Park, CA.

Hehre, W. J.; Ditchfield, R.; Radom, L.; Pople, J. A. (1970). Molecular orbital theory of the electronic structure of organic compounds. V. Molecular theory of bond separation, *J. Am. Chem. Soc.*, 92, pp. 4796–4801.

Hehre, W. J. (1995). *Practical Strategies for Electronic Structure Calculations*, Wavefunction, Irvine, CA.

Hehre, W. J.; Radom, L.; Schleyer, P. von. R.; Pople, J. A. (1986). Ab Initio *Molecular Orbital Theory*, Wiley, New York.

Heitler, W.; London, F. (1927). Wechselwirkung neutraler Atome und homöopolare Bindung nach der Quantenmechanik, *Zeit. f. Phys.*, 44, pp. 455–472.

Hoffmann, R. (1963). An extended Hückel theory I. hydrocarbons, *J. Chem. Phys.*, 39, pp. 1397–1412.

House, J. E. (1998). *Fundamentals of Quantum Mechanics*, Academic Press, San Diego, CA.

Hückel, E. (1931). Quantentheoretische Beiträge zum Benzolproblem, *Zeit. f. Physik*, 70, pp. 204–286 and subsequent papers to 1933.

Irikura, K. K. (1998). Essential Statistical Thermodynamics in Computational Thermochemistry. In Irikura, K. K.; Furrip, D. J. eds., *Computational*

Thermochemistry, ACS Symposium Series 677, American Chemical Society, Washington D. C.

James, H. M.; Cooledge, A. S. (1933). The Ground State of the Hydrogen Molecule, *J. Chem. Phys.*, 1, pp. 825–835.

Jarowski, P. D.; Wodrich, M. D.; Wannere, C. S.; Schleyer, P. von R.; Houk, K. N. (2004). How large is the conjugative stabilization of diynes?, *J. Am. Chem. Soc.*, 126, pp. 15036–15037.

Jensen, F. (1999). *Introduction to Computational Chemistry*, Wiley, New York.

Jensen, J. L. (1976). Heats of hydrogenation: a brief summary, *Progress in Physical Organic Chemistry*, 12, pp. 189–228.

Jurs, P. C., (1996). *Computer Software Applications in Chemistry*, 2nd ed. Wiley, New York.

Kauzmann, W. (1966). *Kinetic Theory of Gases*, Benjamin, New York.

Kistiakowsky, G. B.; Romeyn, Jr., J. R.; Smith, H. A.; Vaughan, W. E. (1935a). Heats of organic reactions. I. The apparatus and the heat of hydrogenation of ethylene, *J. Am. Chem. Soc.*, 57, pp. 65–75.

Kistiakowsky, G. B.; Ruhoff, J. R.; Smith, H. A.; Vaughan, W. E. (1935b). Heats of organic reactions. II. Hydrogenation of some simpler olefinic hydrocarbons, *J. Am. Chem. Soc.*, 57, pp. 876–882.

Kistiakowsky, G. B.; Ruhoff, J. R.; Smith, H. A.; Vaughan, W. E. (1936a). Heats of organic reactions. III. Hydrogenation of some higher olefins, *J. Am. Chem. Soc.*, 58, pp. 137–145.

Kistiakowsky, G. B.; Ruhoff, J. R.; Smith, H. A.; Vaughan, W. E. (1936b). Heats of organic reactions. IV. Hydrogenation of some dienes and of benzene, *J. Am. Chem. Soc.*, 58, pp. 146–153.

Kistiakowsky, G. B.; Nickle, A. G. (1951). Ethane-ethylene and propane-propylene equilibria, *Disc. Faraday Soc.*, pp. 175–187.

Klotz, I. M.; Rosenberg, R. M. (2000). *Chemical Thermodynamics. Basic theory and methods*, 6th ed. Wiley Interscience, New York.

Kreyszig, E. (1988). *Advanced Engineering Mathematics*, 6th ed. Wiley, New York.

Laidler, K. J.; Meiser, J. H. (1999). *Physical Chemistry*, Houghton Mifflin Co. Boston.

Larsen, D. (1975). *Microcomputer Interfacing Experiments*, E & L Instrument Co., 70 Fulton Terrace, New Haven, CT 06512.

Lee, C; Yang, W.; Parr, R. G. (1988). Development of the Colle-Salvetti correlation-energy formula into a functional of the electron density, *Phys. Rev. B*, 37, pp. 785–789.

Levine, I. N. (2000). *Quantum Chemistry*, 5th ed. Prentice-Hall, Upper Saddle River, NJ.

Lewis, G. N.; Randall, M.; Pitzer, K. S.; Brewer, L. (1961). *Thermodynamics*, 2nd ed. Mc Graw-Hill, New York.

Li, Z.; Rogers, D. W.; McLafferty; F. J.; Mandziuk, M.; Podosenin, A. V. (1999). Ab initio calculations of enthalpies of hydrogenation, isomerization, and formation of cyclic C6 hydrocarbons. Benzene isomers, *J. Phys. Chem. A*, 103, pp. 426–430.

Lii, J.-H.; Allinger, N. L. (1989). Molecular mechanics. The MM3 force field for hydrocarbons. 3. The van der Waals potentials and crystal data for aliphatic and aromatic hydrocarbons, *J. Am. Chem. Soc.*, 111, pp. 8576–8582.

Lowe, J. P. (1993). *Quantum Chemistry*, 2nd ed. Academic Press, San Diego, CA.

Martin, J. M. L. (1998). Calibration study of atomization energies of small polyatomics. In Irikura, K. K.; Furrip, D. J. eds., *Computational Thermochemistry*, ACS Symposium Series 677, American Chemical Society, Washington D. C.

Mc Quarrie, D. A. (1983). *Quantum Chemistry*, University Science Books, Sausalito, CA.

Mc Quarrie, D. A.; Simon, J. D. (1999). *Molecular Thermodynamics*, University Science Books, Sausalito, CA.

McWeeny, R. (1979). *Coulson's Valence*, Oxford Univ. Press, Oxford.

Millikan, R. S.; Rieke, C. A.; Orloff, D.; Orloff, H. (1949). Formulas and numerical tables for overlap integrals, *J. Chem. Phys.*, 17, pp. 1248–1267.

Møeller, C.; Plesset, M. S. (1934). Note on an approximation treatment for many-electron systems, *Phys Rev.*, 46, pp. 618–622.

Montgomery, J. A.; Ochterski; J. W.; Petersson, G. A. (1994). A complete basis set model chemistry. IV. An improved atomic pair natural orbital method, *J. Chem. Phys*, 101, pp. 5900–5909.

Mortimer, R. G., (1999). *Mathematics for Physical Chemistry*, 2nd ed. Academic Press, San Diego, CA .

Murrell, J. M., Kettle, S. F. A. and Tedder, J. M. (1985). *The Chemical Bond*, 2nd ed. Wiley, New York.

Nevins, N.; Chen, K.; Allinger, N. L. (1996a). Molecular mechanics (MM4) calculations on alkenes, *J. Comp. Chem.*, 17, pp. 669–694.

Nevins, N.; Lii, J-H.; Allinger, N. L. (1996b). Molecular mechanics (MM4) calculations on conjugated hydrocarbons, *J. Comp. Chem.*, 17, pp. 695–729.

Nyden, M. R.; Petersson, G. A. (1981). Complete basis set energies. I. The asymptotic convergence of pairnatural orbital expansions, *J. Chem. Phys.*, 75, pp. 1843–1862.

Pan, J.-W.; Rogers, D. W.; Mc Lafferty, F. J. (1999). Density functional calculations of enthalpies of hydrogenation, isomerization, and formation of small cyclic hydrocarbons, *J. Molecular Structure (Theochem.)* , 468, pp. 59–66.

Pariser, R.; Parr, R. G., (1953a). A semi-empirical theory of the electronic spectra and electronic structure of complex unsaturated molecules. I., *J. Chem. Phys.*, 21, pp. 466–471.

Pariser, R.; Parr, R. G., (1953b). A semi-empirical theory of the electronic spectra and electronic structure of complex unsaturated molecules. II., *J. Chem. Phys.*, 21, pp. 767–776.

Parr, R. G., (1963). *Quantum Theory of Molecular Structure*, W. A. Benjamin, New York.

Parr, R. G.; Yang, W. (1989). *Density Functional Theory of Atoms and Molecules*, Oxford Univ. Press, New York.

Pauling, L.; Wheland, G. W. (1933a). The Nature of the Chemical Bond. V. The Quantum-Mechanical Calculation of the Resonance Energy of Benzene and Naphthalene and the Hydrocarbon Free Radicals, *J. Chem. Phys.*, 1, pp. 362–374.

Pauling, L.; Sherman, J. (1933b). The Nature of the Chemical Bond. VI. The Calculation from Thermochemical Data of the Energy of Resonance of Molecules Among Several Electronic Structures, *J. Chem. Phys.*, 1, pp. 606–617.

Pauling, L.; Wilson, E. B. (1935). *Introduction to Quantum Mechanics*, Mc Graw-Hill, New York. Reprinted (1963). Dover, New York.

Pauling. L. (1960). *The Nature of the Chemical Bond*, Cornell Univ. Press, Ithaca.

Pedley, J. B.; Naylor, R. D.; Kirby, S. P. (1986). *Thermochemical Data of Organic Compounds*, 2nd ed. Chapman and Hall, London.

Petersson, G. A.; Al-Laham M. A. (1991a). A complete basis set model chemistry. II. Open-shell systems and the total energies of the first-row atoms, *J. Chem. Phys.*, 94, pp. 6081–6090.

Petersson, G. A.; Tensfeldt, T.; Montgomery, J. A. (1991b). A complete basis set model chemistry. III. The complete basis set-quadratic configuration interaction family of methods, *J. Chem. Phys*, 94, pp. 6091–6101.

Pople, J. A. (1953). Electron interaction in unsaturated hydrocarbons, *Trans. Faraday Soc.*, 49, pp. 1375–1385.

Pople, J. A., Head-Gordon, M.; Fox, D. J.; Raghavachari, K.; Curtiss, L. A. (1989). Gaussian-1 theory: A general procedure for prediction of molecular energies, *J. Chem. Phys.*, 90, pp. 5622–5629 and numerous papers in this series.

Pople, J. A. (1999). Nobel lectures. Quantum chemical models, *Rev. Mod. Phys.*, 71, pp. 1267–1274.

Prosen, E. J.; Maron, F. W.; Rossini, F. D. (1951). Heats of combustion, formation, and insomerization of ten C_4 hydrocarbons, *J. Res. NBS.*, 46, pp. 106–112.

Pullman, A.; Pullman, B. (1955). *Cancerisation par les Substances Chimiques et Structure Moleculaire*, Masson. Paris.

Pullman, B. (1962). *The Modern Theory of Molecular Structure* (Engl. Ed.), Dover Pub. Co. New York.

Raghavachari, K.; Stefanov, B. B.; Curtiss. L. A. (1997). Gaussian-1 theory: A general procedure for prediction of molecular energies, *J. Chem. Phys.*, 106, pp. 6764–6767.

Ratner, M. A.; Schatz, G. C. (2001). *Introduction to Quantum Mechanics in Chemistry*, Prentice-Hall, Upper Saddle River, NJ.

Rice, J. R. (1983). *Numerical Methods, Software and Analysis*, Mc Graw-Hill, New York.

Rioux, F. (1987). A simple self-consistent field calculation for two-electron systems, *European Journal of Physics*, 8, pp. 297–299.

Rogers, D. W.; Mc Lafferty, F. J. (1971). A new hydrogen microcalorimeter. Heats of hydrogenation of allyl and vinyl unsaturation adjacent to a ring, *Tetrahedron*, 27, pp. 3765–3775.

Rogers, D. W. (1973). Direct-injection enthalpimetry of micromolar quantities of unsaturated fatty acids, *Analytical Biochem.*, 56, pp. 460–464.

Rogers, D. W.; von Voithenberg, H.; Allinger, N. L. (1978). Heats of hydrogenation of the *cis* and *trans* isomers of cyclooctene, *J. Org. Chem.*, 43, pp. 360–361.

Rogers, D. W.; Dagdagan, O. A.; Allinger, N. L. (1979). Heats of hydrogenation and formation of linear alkynes and a molecular mechanics interpretation, *J. Am. Chem. Soc.*, 101, pp. 671–676.

Rogers, D. w.; Choi, L. S.; Beauregaard, J. A. (1980). An UJT-controlled pulse microcalorimeter heating circuit, *Mikrochim. Acta*, I, pp. 245–258.

Rogers, D. W.; Crooks, E. L. (1983). Enthalpies of hydrogenation of the isomers of *n*-hexene, *J. Chem. Thermodynam.*, 15, pp. 1087–1092.

Rogers, D. W.; Dejroongruang, K. (1988). Enthalpies of hydrogenation of the *n*-heptenes and the methylhexenes, *J. Chem. Thermodynam.*, 20, pp. 675–680.

Rogers, D. W.; Mc Lafferty F. J.; Podosenin, A. V. (1996). *Ab initio* calculations of enthalpies of hydrogenation and isomerization of cyclic C4 hydrocarbons, *J. Phys Chem.*, 100, pp. 17148–17151.

Rogers, D. W.; Mc Lafferty F. J.; Podosenin, A. V. (1998a). G2 *ab initio* calculations of enthalpies of hydrogenation, isomerization, and formation of C3 cyclic ketones and alcohols, *J. Org. Chem.*, 63, pp. 7319–7321.

Rogers, D. W.; Zhao, Y.; Traetteberg, M.; Hulce, M.; Liebman, J. F. (1998b). Enthalpies of hydrogenation and formation of enones. Resonance energies of 2-cyclopentenone and 2-cyclohexenone, *J. Chem. Thermodynam.*, 30, pp. 1393–1400.

Rogers, D. W.; McLafferty, F. J. (2001). The influence of substituent groups on the resonance stabilization of benzene. An *ab initio* computational study, *J. Org. Chem.*, 66, pp. 1157–1162.

Rogers, D. W. (2003). *Computational Chemistry for the PC*, 3rd ed. Wiley. New York.

Rogers, D. W.; Matsunaga, N. McLafferty, F. J.; Zavitsas, A. A.; Liebman, J. F. (2004). On the lack of conjugation stabilization in polyynes (polyacetylenes), *J. Org. Chem.*, 69, pp. 7143–7147.

Rogers, D. W.; Matsunaga, N.; Zavitsas, A. A. (2005). G3(MP2) enthalpies of hydrogenation isomerization, and formation of extended linear polyacetylenes, *J. Phys. Chem. A*, 109, pp. 9169–9173.

Rony, P. (1976). *Bugbook V. Introductory Experiments in Digital electronics*, E & L Instrument Co., 70 Fulton Terrace, New Haven, CT 06512.

Roothaan, C. C. J. (1951). New developments in molecular orbital theory, *Rev. Mod. Phys.*, 23, pp. 69–89.

Rosen, N. (1931). The normal state of the hydrogen molecule, *Phys. Rev.*, 38, pp. 2099–2114.

Roth, W. R.; Bang, W. B.; Goebel, P.; Sass, R. L.; Turner, R. B. (1964). On the Question of Homoconjugation in *cis,cis,cis*-1,4,7-Cyclononatriene, *J. Am. Chem. Soc.*, 86, pp. 3178–3179.

Roth, W. R.; Lennartz, H.-W. (1980a). Hydrierwärmen, I. Bestimmung von Hydrierwärmen mit einem isothermen Titrationskalorimeter, *Chem. Ber.*, 113, pp. 1806–1817.

Roth, W. R.; Klaerner F.-G.; Lennartz, H.-W. (1980b). Hydrierwärmen, II. Hydrierwärme des Bicyclo[2.1.0]pent-2-ens, ein antiaromatiches System II, *Chem. Ber.*, 113, pp. 1818–1829.

Roth, W. R.; Scholz, B. P. (1981). Das Energieprofil des o-Chinodimethan ⇌ Benzocyclobuten-Gleichgewichtes, II, *Chem Ber.*, 114, pp. 3741–3750.

Roth, W. R.; Scholz, B. P.; Breuckmann, R.; Jelich, K.; Lennartz, H-W. (1982a). Thermolyse des 1,2,6,7-Octatetraens, *Chem. Ber.*, 115, pp. 1934–1946.

Roth, W. R.; Kirmse, W.; Hoffmann, W.; Lennartz, H.-W. (1982b). Hydrierwärmen, III. Einfluss von Fluorsubstituenten auf die termische Umlagurung des Cyclopropansystems III, *Chem. Ber.*, 115, pp. 2508–2515.

Roth, W. R.; Klarner, F-G.; Grimme, W.; Koser, H.; Busch, R; Muskulus, B.; Breuckmann, R.; Scholtz, B. P.; Lennartz, H-W. (1983). Stereochemistry of the bicyclo[2.1.0]pentane ring opening; thermolysis of tricyclo[3.2.0.0^{2,4}]heptane derivatives, *Chem Ber.*, 116, pp. 2717–2737.

Roth, W. R.; Lennartz, H-W.; Vogel, E.; Leiendecker, M.; Oda, M. (1986). Resonanzenergie kondensierte [4n]Annulene, *Chem. Ber.*, 119, pp, 837–843.

Roth, W. R.; Langer, M. B.; Stevermann, B.; Maier,G.; Reisenauer, H-P; Sustmann. R.; Muller, W. (1987). The Diradical 2,3,5,6-Tetramethylene-1,4-cyclohexanediyl ("1,2,4,5-Tetramethylene-benzene"), *Angew. Chem. Int. Ed. Engl.*, 26 (3), pp. 256–258.

Roth, W. R.; Lennartz, H.-W.; Doering, von E.; Dolbier Jr., W. R.; Schmidhauser, J. C.; (1988). Thermochemistry of the orthogonal butadienes (Z,Z)-3,4-dimethylhexa-2,4-diene and 2,3-di-*tert*-butylbuta-1,3-diene, *J. Am. Chem. Soc.*, 110, pp. 1883–1889.

Roth, W. R.; Bauer, F.; Braun, K.; Offerhaus, R. (1989). Energy wells of diradicals; 4-methylene-1,3-cyclopentanediyl, *Angew. Chem. Int. Ed. Engl.*, 28, (8), pp. 1056–1057.

Roth, W. R.; Lennartz, H.-W.; Doering, W. von E.; Birladeanu, L; Guyton, C. A.; Kitagawa, T. (1990). A "frustrated" Cope rearrangement: thermal interconversion of 2,6-diphenylhepta-1,6-diene and 1,5-diphenylbicyclo[3.2.0]heptane, *J. Am. Chem. Soc.*, 112, pp. 1722–1732.

Roth, W. R.; Ruhkamp, J.; Lennartz, H-W. (1991a). Bestimmung der Singulet-Triplett-Aufspaltung von Diradikalen mit Hilfe der Sauerstoff-Abfang-Technik, *Chem. Ber.*, 124, pp. 2047–2051.

Roth, W. R.; Adamczak, O.; Breuckman, R.; Lennartz, H.-W.; Boese, R. (1991b). Die Berechnung von Resonanzenergien; das MM2ERW Kraftfeld, *Chem. Ber.*, 124, pp. 2499–2521.

Roth, W. R.; Langer, R.; Ebbrecht, T.; Beitat, A.; Lennartz, H.-W. (1991c). Zur Energie-Delle von Diradikalen, III. 2,3,5,6-Tetramethylen-1,4-cyclohexadiyl, *Chem. Ber.*, 124, pp. 2751–2760.

Roth, W. R.; Klaerner, F.-G.; Hydrierwärmen, IV. Siepert, G.; Lennartz, H.-W. (1992). Zur Frage der Homoaromatizität von Norcaradien und Cycloheptatrien, *Chem. Ber.*, 125, pp. 217–224.

Roth, W. R.; Wollweber, D.; Offerhaus, R.; Rekowski, V.; Lennartz, H-W.; Sustmann, R.; Muller, W. (1993a). 2-Methylen-1,4-cyclohexadiyl, *Chem. Ber.*, 126, pp. 2701–2715.

Roth, W. R; Winzer, M.; Lennartz, H.-W.; Boese R. (1993b). Zur Energiedelle des orthogonallen – Trimethylenmethans. 1-Methylen-2-phenylcyclopropan- Thermolyse, *Chem Ber.*, 126, pp. 2717–2725.

Roth, W. R.; Hopf, H.; Horn, C. (1994). Propargyl-Stabilisierungsenergie, *Chem. Ber.*, 127, pp. 1781–1795.

Roth, W. R.; Wildt, H.; Schlemenat, A. (2001). Trimethylenemethane derivatives stabilized by conjugation II – concerted or nonconcerted generation? *Eur. J. Org. Chem.*, pp. 4081–4099.

Sabbe, M. K.; Saeys, M.; Reyniers, M-F.; Marin, G. B.; van Speybroeck V.; Waroquier, M. (2005). Group additive values for the gas phase standard enthalpy of formation of hydrocarbons and hydrocarbon radicals, *J. Phys. Chem. A*, 109, pp. 7466–7480.

Salem, L. (1966). *The Molecular Orbital Theory of Conjugated Systems*, Benjamin, New York.

Saunders, M. (1987). Stochastic exploration of molecular mechanics energy surfaces. Hunting for the global minimum, *J. Am. Chem. Soc.*, 109, pp. 3150–3152.

Scheid, F. (1968). *Numerical Analysis*, Schaum's, Mc Graw-Hill, New York.

Schlecht, M. F. (1998). *Molecular Modeling on the PC*, Wiley-VCH, New York.

Schmidt, J. C.; Gordon, M. S. (1998). The construction and interpretation of MCSCF wavefunctions, *Ann. Rev. Phys. Chem.*, 49, 233–266.

Schmidt, M. W.; Baldridge, K. K.; Boatz, J. A.; Elbert, S. T.; Gordon, M. S.; Jensen, J. H.; Koseki, S.; Matsunaga, N.; Nugyen, K. A.; Su, S.; Windus, T. L.; Dupuis, M.; Montgomery, J. A. (1993). General atomic and molecular electronic structure system, *J. Comp. Chem.*, 14, pp. 1347–1363.

Schroedinger, E. (1926). Quantisierung als Eigenwertproblem, *Ann. der Phys.*, 79, pp. 361–376 and subsequent papers in 1926.

Schroedinger, E. (1928a). *Abhandlungen zur Wellenmechanik*, Barth, Leipzig.

Schroedinger, E. (1928b). *Collected Papers on Wave Mechanics*, Blackie & Son, London & Glasgow.

Schwartz, J. T. (1961). *Introduction to Matrices and Vectors*, Dover, NY.

Scott A. P.; Radom, L. (1996). Harmonic vibrational frequencies: an evaluation of Hartree-Fock, Møeller-Plesset, quadratic configuration interaction, density

functional theory, and semiempirical scale factors, *J. Phys Chem.*, 100, pp. 16502–16513.
Skinner, H. A.; Snelson, A. (1959). Heats of hydrogenation Part 3, *Trans. Faraday Soc.*, 55, pp. 404–407, see also Parts 1 and 2.
Slater, J. C. (1931). Molecular energy levels and valence bonds, *Phys. Review.*, 38, pp. 1109–1144.
Smith, W. B. (1996). *Introduction to Theoretical Organic Chemistry and Molecular Modeling*, VCH, New York.
Starzak, M. E. (1989). *Mathematical Methods in Chemistry and Physics*, Plenum, New York.
Steele, W. V.; Chirico, R. D.; Cowell, A. B.; Knipmeyer, S. E.; Nguyen, A. (2002). Thermodynamic properties and ideal-gas enthalpies of formation for *trans*-methyl cinnamate, α-Methyl cinnamaldehyde, methyl methacrylate, 1-nonyne, trimethylacetic acid, trimethylacetic anhydride, and ethyl trimethylacetate, *J. Chem. Eng. Data*, 47, pp. 700–714.
Steiner, E. (1996). *The Chemistry Maths Book*, Oxford Sci., New York
Stewart, J. J. P. (1989). Optimization of parameters for semiempirical methods I. Method, *J. Comp. Chem.*, 10, 209–220.
Stewart, J. J. P. (1990). MOPAC: A general molecular orbital package, *J. Comp. Aided Molecular Design*, 4, 1–105.
Streitwieser, A. (1961). *Molecular Orbital Theory for Chemists*, Wiley, New York.
Stull, D. R.; Westrum, E. F.; Sinke, G. C. (1969). *The Chemical Thermodynamics of Organic Compounds*, Wiley, NY.
Thiel, W. (1998). Thermochemistry from Semiempirical MO Theory. In Irikura, K. K.; Furrip, D. J. eds., *Computational Thermochemistry*, ACS Symposium 677, American Chemical Society, Washington D. C.
Turner, R. B.; Garner, R. H. (1957). Heats of Hydrogenation. V. Relative stabilities in certain exocyclic-endocyclic olefin pairs, *J. Am. Chem. Soc.*, 80, pp. 1424–1430.
Turner, R. B.; Mallon, B. J.; Tichy, M.; Doering, W. von E.; Roth. W. R.; Schroeder, G. (1973). Heats of hydrogenation. X. Conjugative interaction in cyclic dienes and trienes, *J. Am. Chem. Soc.*, 95, pp. 8605–8610.
van Hemelrijk, D. V.; van den Enden, L.; Geise, H. J.; Sellers, H. L.; Schafer, L. (1980). Structure determination of 1-butene by gas electron diffraction, microwave spectroscopy, molecular mechanics, and molecular orbital constrained electron diffraction, *J. Am. Chem. Soc.*, 102, pp. 2189–2195.
Wheland, G. W. (1955). *Resonance in Organic Chemistry*, Wiley, New York.
Wiberg, K. B.; Fenoglio, R. A. (1968). Heats of formation of C4H6 hydrocarbons *J. Am. Chem. Soc.*, 90, pp. 3395–3397.
Wiberg, K. B.; Crocker, L. S.; Morgan, K. M. (1991). Thermochemical studies of carbonyl compounds. 5. Enthalpies of reduction of carbonyl groups *J. Am. Chem. Soc.*, 113, pp. 3447–3450.

Young, D. C. (2001). *Computational Chemistry: A Practical Guide for Applying Techniques to Real World Problems*, Wiley, New York.

Young, H. D. (1962). *Statistical Treatment of Experimental Data*, Mc Graw-Hill, New York.

Young, H. D.; Friedman, R. A. (2000). *Sears and Zemansky's University Physics*, 10th ed. Addison Wesley Longmans, San Francisco, CA.

Zavitsas, A. A. (2001). Logic *vs.* misconceptions in undergraduate organic textbooks: radical stabilities and bond dissociation energies, *J. Chem. Ed.*, 78, pp. 417–419.

Miscellaneous ancillary references:

PCMODEL v. 8.0 (1993–2003). Serena Software, Box 3076, Bloomington IN 47402-3076.

Sigma Plot for Windows 5.0, 8.0 SPSS Inc. 1986–1999.

TableCurve Windows v 1.0 User's Manual, (1992). AISN Software, Jandel Scientific, San Rafiel, CA.

TableCurve, Jandel Scientific, SYSTAT Software Inc., 510 Canal Boulevard, Suite F, Richmond CA.

Index

1–4 interaction, 165
1–5 interaction, 165
1,3-butadiene, 3
1-heptene, 13

ab initio, 154, 176
ab initio method, 182
accuracy, 15
acetaldehyde, 132
acetone, 132
acetylene, 23
acids, 137–147
A-D converter, 21
additivity, 159
alcohols. 196
alkane, 9
alkanes, 169
alkene, 3, 9, 165, 169
alkynes, 3, 7 165
Allinger, 154, 169
AM1, 181
antibonding, 180
applications, 15
aromaticity, 1
atomization method, 189
atomization, 163
azides, 149

B3LYP, 200, 202
basis function, 174
basis set, split valence, 184
BDE, 202
Becke, 200
Benson, 154

benzene, 5, 15, 16
bomb calorimetric, 167
bond additivity, 154, 159
bond dissociation energy, 202
bond energies, 160
bonding, 180
Born, 174
Boys, 183
Burkert, 170
butanes, 26 $f\!f$
butenal, 133

cartesian coordinates, 155
CBS, 186, 201
cholestenes, 131
CI, 186
cis-2-butene, 171
CNDO, 181
combustion thermochemistry, 4, 6
commercial instruments, 17
complete basis set, 201
computational chemistry, 5
conjugation, 5
contracted Gaussian, 184
core, 177
correlation energy, 185, 197, 200
Curtiss, 188
cycloheptatriene, 10
cycloheptatrienilyum bromide, 151
cycloheptatrienilyum chloride, 151

decenes, 100 $f\!f$, 108
definitions, 1
density functional theory, 199

Dewar, 154, 175, 181
DFT, 199
dihedral angle, 156
dipole moment, 157
dissociation, 160
dodecenes, 113–119

eicosene, 130
electron affinity, 197
empirical methods, 166
energy, 1, 161
enthalpy, 1, 161
enthalpy of
    atomization, 163
    formation, 2
    fusion, 13
    hydrogenation, 172, 193
    solution, 13
    transfer, 12
    vaporization, 13
entropy, 169
equilibrium configuration, 170
equilibrium constant, 169
esters, 143–149
ethene, 9, 24, 164, 171, 178, 193
ethers, 133–145
ethyne, 23
exchange correlation, 200
exchange energy, 200
experimental results, 23 *ff*

fatty acids and esters, 143–149
first ionization potential, 197
first law, 4
flow calorimeter, 8
fluorides, 150–151
force field, 168
free radical, 202
functional, 199

G3(MP2), 186, 189
G3(MP2)B3, 202
GAMESS, 186

gas phase, 12
gauche interaction, 165
Gaussian, 154
Gaussian basis set, 183
geometry, 155
Gibbs free energy, 169
glacial acetic acid, 9
global minimum, 170
Gordon, 173, 176
graphical user interface, 171
ground state energies, 168
group additivity, 154, 163
GTO, 183

Hamiltonian operator, 172
Hartree–Fock, 198
Hartree–Fock limit, 185
heat capacity, 9, 169, 191
heat of
    activation, 14
    combustion, 2, 166
    hydrogenation, 2
Hehre, 193
heptadecenes, 127–128
heptynes, 53 *ff*, 60
heteroatoms, 165
hexadecenes, 125–126
hexenes, 33 *ff*
Hilbert space, 174
history, 1, 6
Hoffmann, 181
homolytic cleavage, 202
Huckel method, 176
hydrocarbons, 2
hydrogen calorimeter, 17
hydrogen thermochemistry, 4
hydrogenation, 195

implementations, 17
indirect calculation, 3
INDO, 181
internal coordinates, 156
ionization potential, 197

isodesmic reaction, 193, 194
isomerization, 4, 171, 195
isomerization enthalpy, 6
isoperibol calorimeter, 10

ketones, 134 – 137, 196
Kistiakowsky, 1, 6, 8, 190, 192, 193
Kohn-Sham equations, 199

latent heats, 13
Lee, 200
linear combination, 174
linear least squares, 22

Mc Lafferty, 10
Methane, 161
Methodology, 8
methyl radical, 203
microburet, 11
microcalorimeter, 10
microcomputer, 10, 21
millivolt, 20
MINDO, 181
miniaturization, 18
Moeller Plesset, 186
molecular mechanics, 154, 168
molecular modeling, 155
molecular orbital, 173
molecular strain, 173
7
NDDO, 181
NDO, 181
nonadecenes, 129
nonenes, 99 $f\!f$
nonynes, 95 $f\!f$, 98
normalization constant, 174

octadecenes, 128 – 129
octenes, 90 $f\!f$
octynes, 74 $f\!f$, 85
oscillatory calorimeters, 10

parameter file, 155
parameterization, 15
Pariser, 181
Parr, 181, 200
partition function, 171
pascal (unit), 2
Pd, 9
pentadecenes, 124–125
pentanes, 27 $f\!f$, 32
perturbation term, 201
PM3, 181
polyalkynes, 153
polyynes, 5
Pople, 173, 176, 181
precision, 14
primitive Gaussian, 184
propargyl radical, 203
propenes, 24 $f\!f$
propenol, 132
propyne, 24
proton affinity, 197, 199
PtO, 10

QCISD(T), 186
quantum mechanics, 5

radicals, 5
resonance stabilization, 5, 16
resonance, 1
Rogers, 1, 7, 10, 195
Roothaan, 175
Roth, 1, 7, 11, 17

safety, 18
SCF, 175
Schroedinger equation, 172
secular determinant, 178, 179
self consistent field, 175
semiempirical methods, 181
semiempirical, 154, 176
Skinner, 1, 7, 9

Slater determinant, 173, 181
solution hydrogenation, 7
solution thermochemistry, 7, 12
solvent effect, 8, 10
solvent interactions, 7
solvent-alkene complex, 15
span, 183
spinorbital, 186
split valence, 184
stability, 17
stabilization, 16
standard state, 2, 153, 162
steric energy, 169
Stewart, 175, 181
STO, 183
STO – 3G, 184

temperture corrections, 14
tert-butylmethane, 194
tetradecenes, 120–124
thermistor, 11, 18
thermochemical cycle, 4, 190
thermochemical database, 166

thermodynamic cycle, 12, 162
thermogram, 22
titration calorimeter, 11
torsional correction, 172
torsional mode, 192
trans-2-butene, 6, 171
tridecenes, 119–120
Turner, 1, 7, 10, 14

undecenes, 110 – 113

variational principle, 174, 177
virtual orbital, 174, 186
visualization, 158

water, 156
wave function, 173
Wiberg, 195
Williams, 1, 7, 12

Yang, 200

z-matrix, 156